Gene Watson Flying Saucers

and

Generalized Propulsion

Book by Stephen Watson

ISBN-10: 061599315X
ISBN-13: 978-0615993157 (human primate press)

About Gene Watson

- He would dumpster dive for material to use in his saucers.

- He would build a flying saucer in the backyard of a house.

- He calls his saucers ground effect vehicles.

- Hi fascination with flying saucers began with the movie "Forbidden Planet."

- He would pick up bear cans and scrap aluminun between jobs.

- During the early 1970s he built a flying saucer in Little Rock, Arkansas with the help of Dr. Bronco, a physics teacher.

- His Arkansas saucer made the newspaper. The FAA issued an order not to fly it over three feet.

- During 1980s American businessman Ross Perot sent a rep to look at his Wylie, Texas flying saucer.

- He likes the American Indian/ Native American.

- His biological father faught against the Nazi's in North Africa.

- He was adopted as a child and grew up near Sherman, Texas.

- He believes in higher moral ethics philosophy.

- He will not design weapons.

- He likes angels and claims to have see an apparition of the Virgin Marry while living in America.

Table Of Contents

About Gene Watson	i
Vintage Photos	ii
Little Rock, Arkansas (early 1970s)	........Page 1
Wylie, Texas (Mid 1980s)	........Page 2
Sachse, Texas (Late 1980s)	........Page 3
Geometric Configuration (50 Years Old)	........Page 4
Shell Cluster Upgrade	........Page 5
Staged Burn Principle	........Page 6
Rocket Array - Hypersonic	........Page 7
Cascade Rocket Burn Motor (30 Years Old)	........Page 8
Particle Accelerator Rocket	........Page 9
Flywheel	........Page 10
Like A Needle Following The Grooves	........Page 11
Motor Storage Device (Explanation)	........Page 12
Why Two?	........Page 13
Toroid Bevatron	........Page 14
What's New About The Toroid Bevatron	........Page 15
Static Thrust	........Page 16
Propulsion Motor Matrix	........Page 17
Propulsion Matrix 3D	........Page 18
Plasma - Ionic Motor	........Page 19
The Jet-Plasma-Ionic-Motor	........Page 20
UFO Motor	........Page 21
Primitive Toroidal Donut	........Page 22
Spiral, Proton Beam	........Page 23
Spaceship Fleet Motor	........Page 24
Chemical Burn Rocket Ice Ball Upgrade	........Page 25
UFO Saucer Dream	........Page 26
Cigar UFO Toroid Bevatron	........Page 27
Cigar UFO Detailed Further	........Page 28
The Nuclear Fusion Toroid Reactor	........Page 29
Primitive UFO Motor	........Page 30
Logical Progression In Propulsion Design	........Page 31
Volts/Meter = Electric Field	........Page 32
Faraday Ion Motor	........Page 33
AC Thrust In Simple Terms	........Page 34
American Flag Electronic Symbol	........Page 35
Buoyant Like Blimp	........Page 36
Perpetual Steady State Kaleidoscope	........Page 37
Electric Field String	........Page 38
Einstein's Fudge Numbers	........Page 39
10,000 B.C.	........Page 40
Chinese Fan	........Page 41
Cherubim Wings	........Page 42

Table Of Contents

Lab: Negative Ion DC Apparatus ....Page 43
Lab: E.L.F. ....Page 44
Lab: AC Thrust Saucers ....Page 45
Lab: Ion Helicopter ....Page 46
Lab: Stack Illustration ....Page 47
Lab: AC Gigar Lighter; Brushless DC Motor ....Page 48
Lab: Oscilloscope ....Page 49
Lab: Circuitry Experiments ....Page 50
Lab: Microwave Detection ....Page 51
Lab: X-rays ....Page 52
Lab: Needle Array ....Page 53
Lab: Wind Bike ....Page 54
Lab Experiment Notes; Arkansas ....Page 55
Wylie, Texas ....Page 56
Sachse; Propulsion Matrix ....Page 57
Inventions Get Stolen ....Page 58
Giving Ideas Away ....Page 59
NASA Rejection; Large Chemical Rockets ....Page 60
Why Mars? ....Page 61
Nuclear Fusion For Space Travel ....Page 62
Fussion vs. Fission; Deuterium & Tritium ....Page 63
UFOs ....Page 64
SETI ....Page 65
Interstellar Spaceship Permanent Home ....Page 66
Ethics ....Page 67
The Secrecy Paradox ....Page 68
Angels; Angel Wings ....Page 69
More on Wylie, Texas ....Page 70
Nuclear "Cold" Fusion Ball ....Page 71
1977 Engine ....Page 72
Nuclear Cold Fusion Deuterium Reactor ....Page 73
No Plasma Cold Fusion ....Page 74
Double Tube Particle Accelerators; Toroid ....Page 75
"Forbidden Planet" Analogy ....Page 76

Glossary ....Page 77
|
Page 92

Scan Documents ....Page 93
|
Page 118

Simple Index ....Page 119
| ....Page 121
References ....Page 122

ACKNOWLEDGMENTS

Thanks to Gene Watson for the sketches and information gathered for this book.

.

And thanks to family members for additional contributions.

Artwork by Stephen Watson

The bell-shaped Wylie, Texas flying saucer under construction.

A flying saucer being built in Garland, Texas sometime during the mid 1980s.

Artist conception of saucer on campus.

Charles Bronco
Assoc. Prof.
Physics

Built and tested in Little Rock Arkansas 1971 with help from Dr. Bronco at Arkansas Polytech College.

Wang Computer calculator
circa the early 1970s

Power: Volvo motor connected to 4 blade 5 foot propeller: gear reduction. It made front of the Arkansas newspaper. Did the motor turn the blades fast? Yes. The Federal Aviation Administration ordered not to make it rise above three feet. This saucer was already 10 foot wide on the campus of Arkansas polytechnic college In Russelville, Arkansas outside the door of the physics building and in the campus yearbook 1970. I built a special tow for it and I started building it under the supervision of Dr. Bronco Arkansas Polytechnic College and finished it in Little Rock Arkansas. I built some of the parts that went into the saucer with the University lathe and he taught me their Wang Computer for calculations.

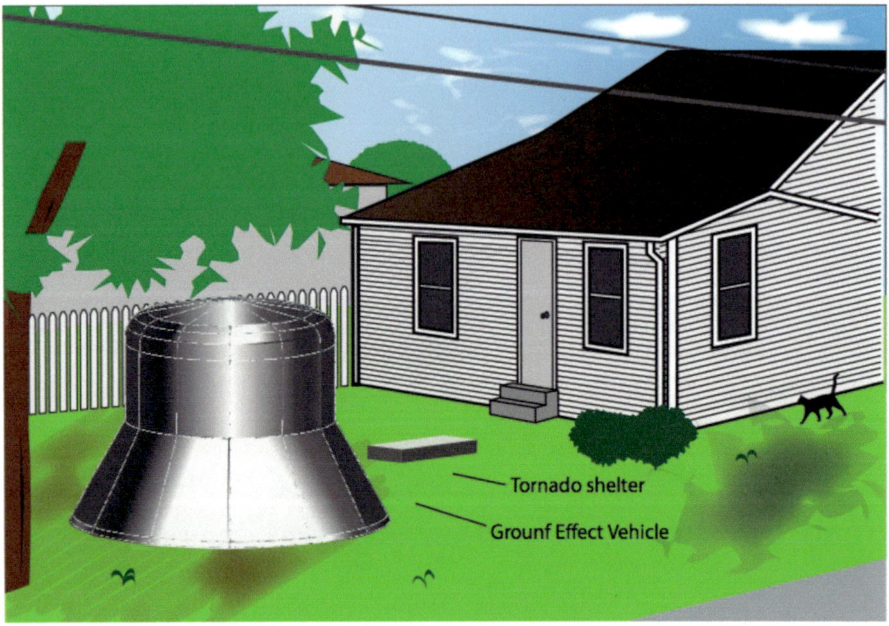

Artist depiction of the 1987 Wylie, Texas flying saucer.

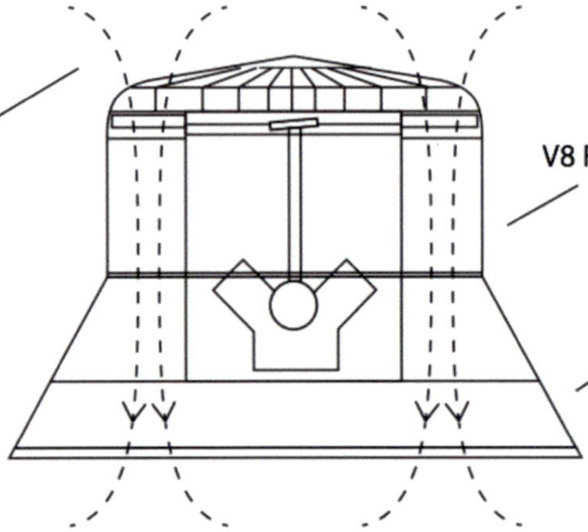

It had a V8 ford car motor. Electric start connected to a gear reduction shaft. The shaft spun. The bell skirt permitted me to take my time learning the controls before I floored the gas pedal to rise past treetop but Wylie, Texas never let me finish building it. It was a ground effect vehicle with the capacity to leave vertically ground effect after the pilot built up enough confidence. Planned ionic thrust implementation.

Artist Depiction. The Sachse, Texas saucer - Late 1980s On Mr. Stone's farm (not completed)

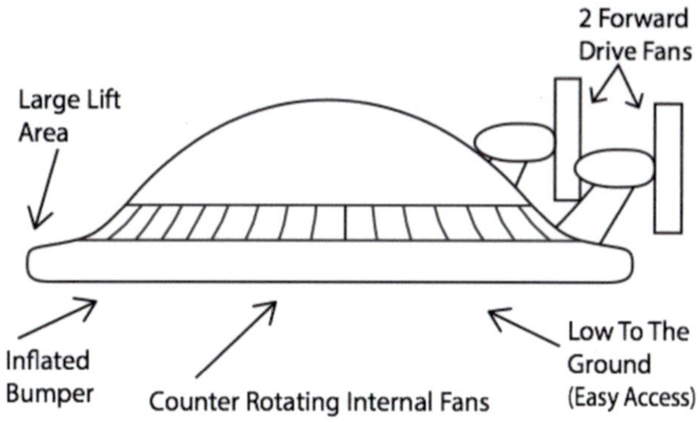

Reason for counter rotating: Centrifugal gyroscopic force cancellation.

Reason for large lift area: Low power to lift ratio saves gas consumption.

Reason for vanes top & bottom: So three limbs wouldn't touch fans.

Reason for infalted bumper: So accidental bumps against objects cause no harm.

THIS IDEA IS 50 YEARS OLD

An example of geometric configuration:

Hydrogen has a flame propogation of about Mach 4. This combines radiation, convection, conduction in afterburner.

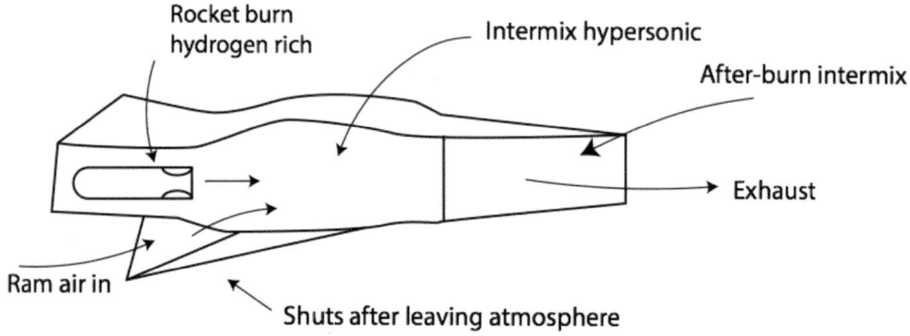

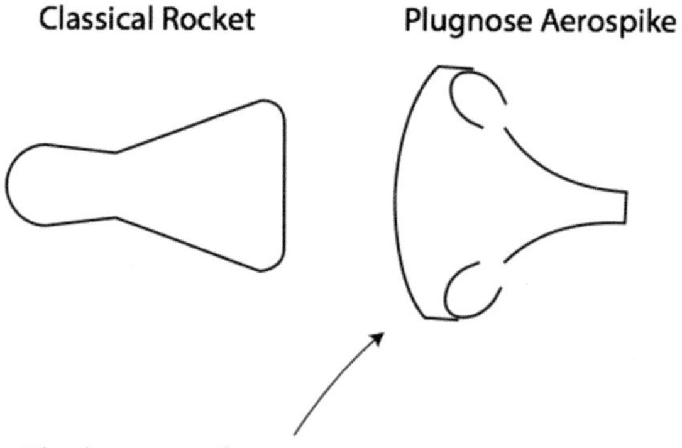

The bottom of my 1962 saucer was aerospike.

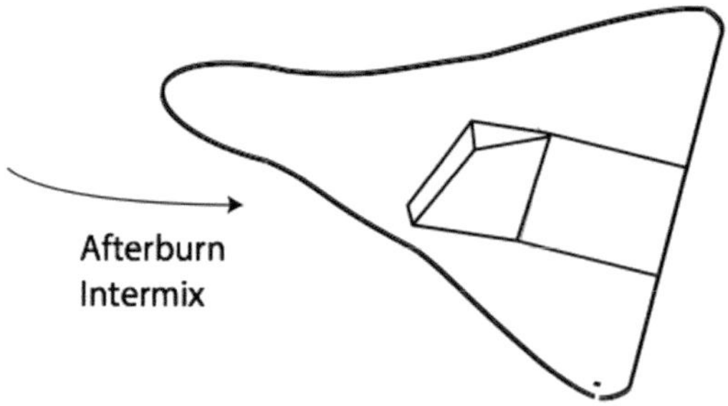

Spaceplane
No staging runway.
Ground to orbital.
Makes the space shuttle obsolete.
Dr. Bronco had this idea at Arkansas Polytechnic College In Russelvill, Arkansas 1970.

THIS IDEA IS 50 YEARS OLD

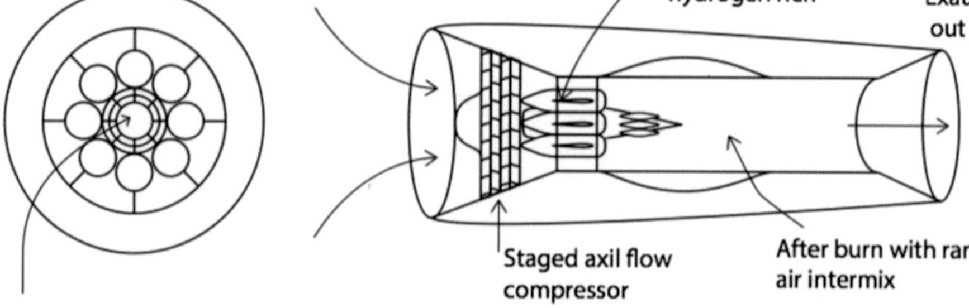

Artillery shell shaped liquid fuel
hydrogen rich burn rocket cluster

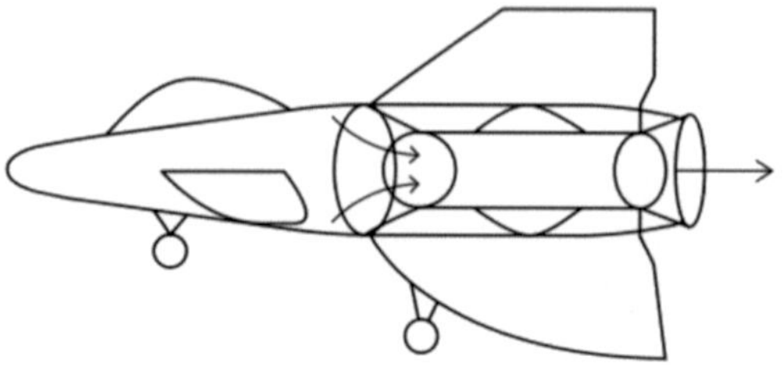

It can take off from a runway
POTENTIALLY GO ORBITAL

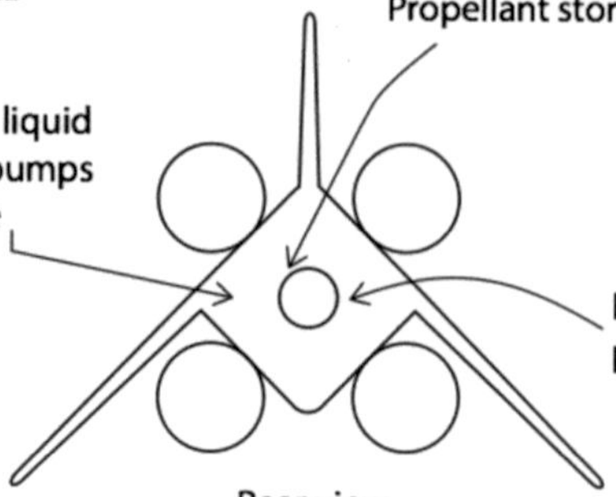

I drew a modification to the pipe bundle rocket so it looks like an artillery shell. This means as a shell cluster bundle it replaces the earlier bundle in the four motor rocket plane. I tried to draw only primitive upgrades to the chemical burn rocket concept. Is it primitive enough for your 21st century civilization?

Staged Burn Principle

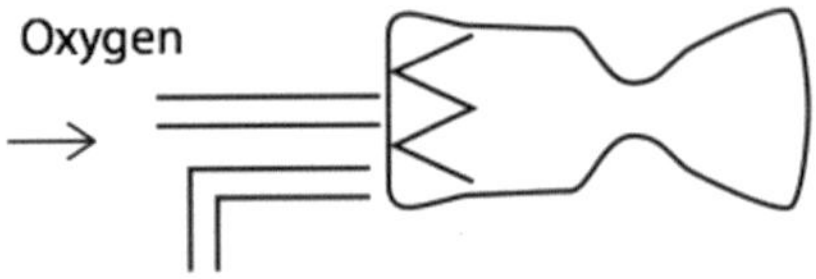

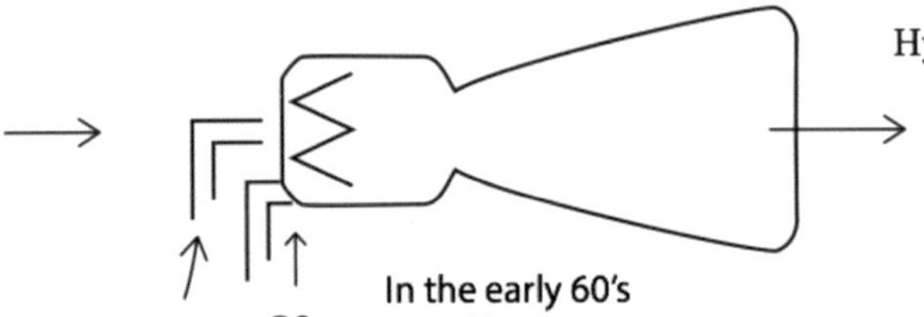

In the early 60's liquid hydrogen as a fuel was perfected.

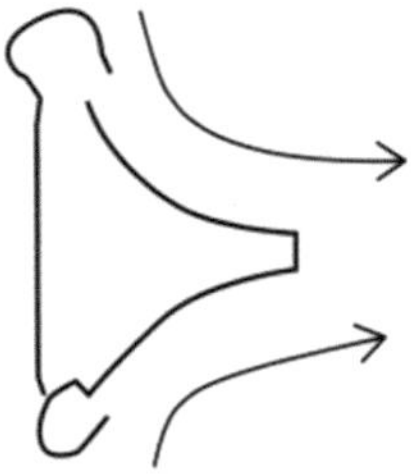

The aerospike sucks in external atmospheric gas resulting in a bigger mass being accelerated.

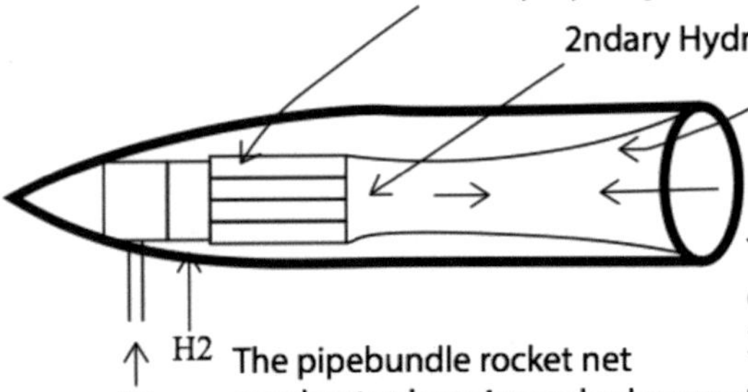

To design this right requires a computer for the mach four 2ndary burn to stay inside rocket.

The pipebundle rocket net accelerates by primary hydrogen rich burn inside the tubes.

Intensify pump pressure to cause higher combustian pressure and a larger nozzle. The throat is a choke that makes the internal burn subsonic.

Dr. Bronco did some calculations and I built a crude device to illustrate the staged burn principle in 1970 (Arkansas Ploytechnic College).

Rocket Array - Hypersonic

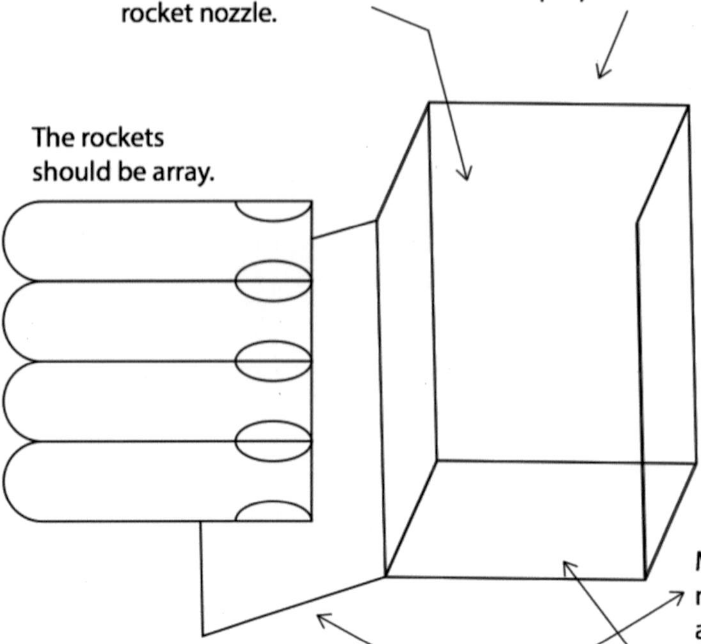

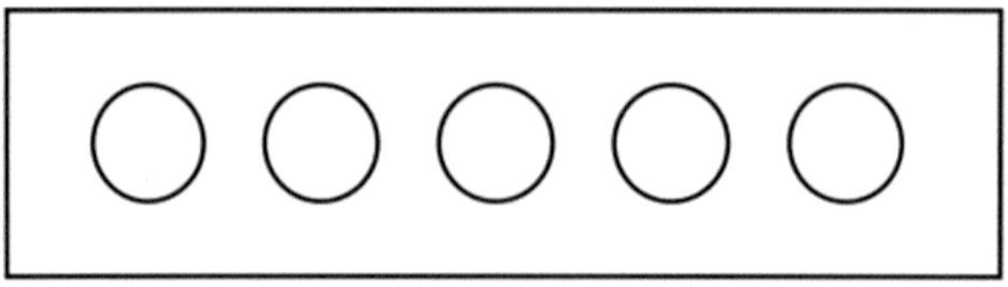

Exhaust

Forget trying to get hypersonic mach 4 internal burn. The intermix is instant. This thing junks the AF secret hypersonic ram jet they haven't got working right yet. The USAF hyperson burn failed because a rocket array was needed. Jet intermix substitute is inferior.

Cascade Burn Motor Design
30 Years Old

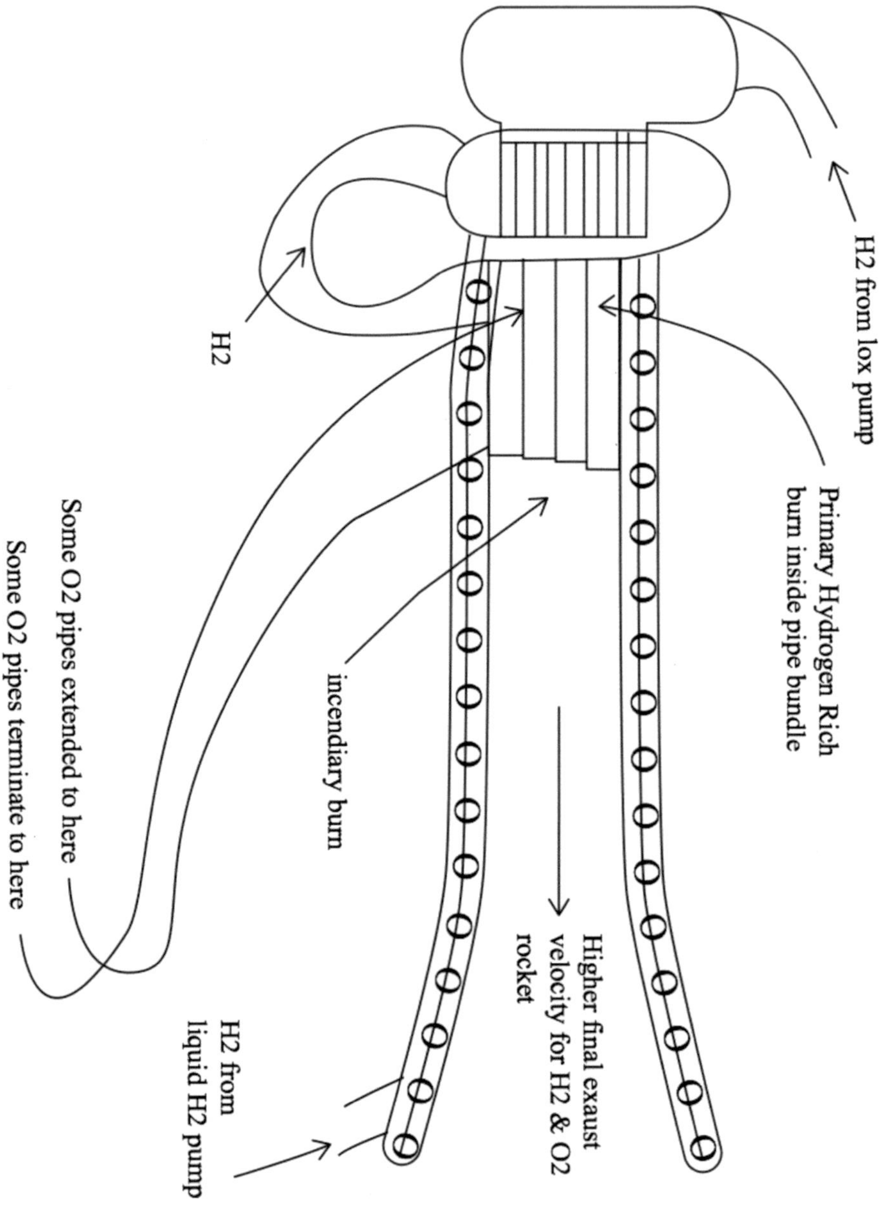

Cascade burn results in a higher final velocity.

Particle Accelerator Rocket

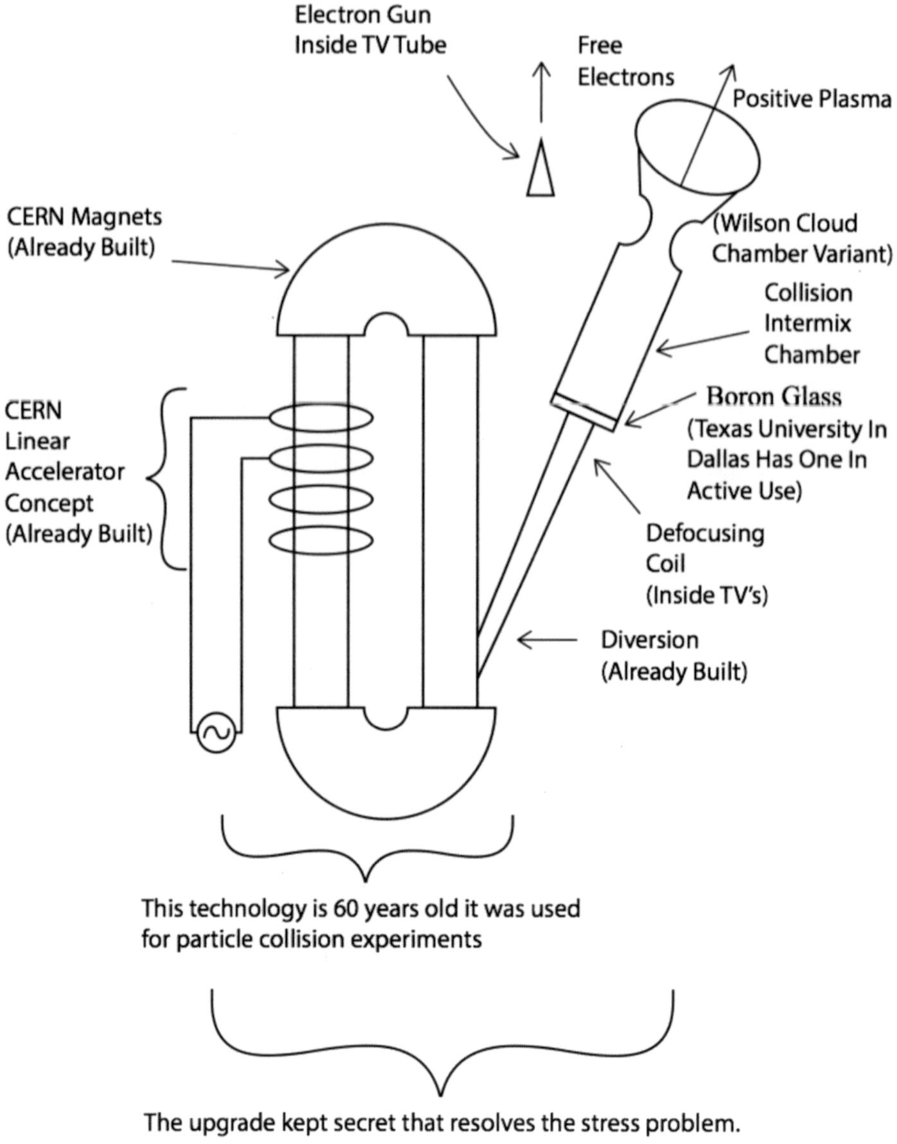

So what did I invent? Nothing. Any physics senior if asked: "Modify a particle accelerator to concentrate thrust as a rocket to generate thrust as a rocket motor" - even the dumbest can detail even better than shown here.

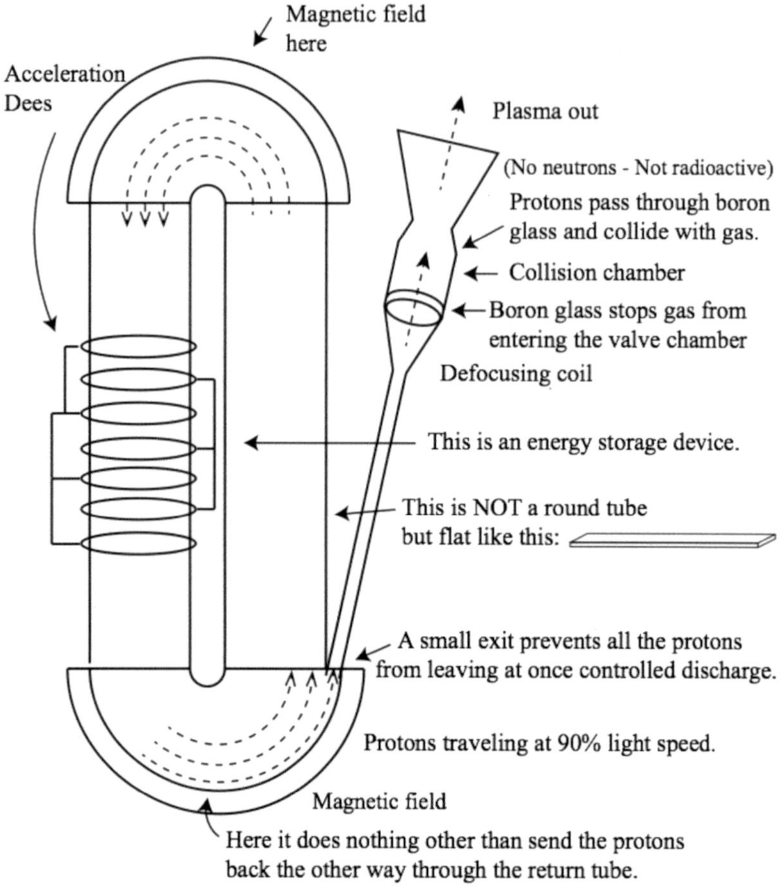

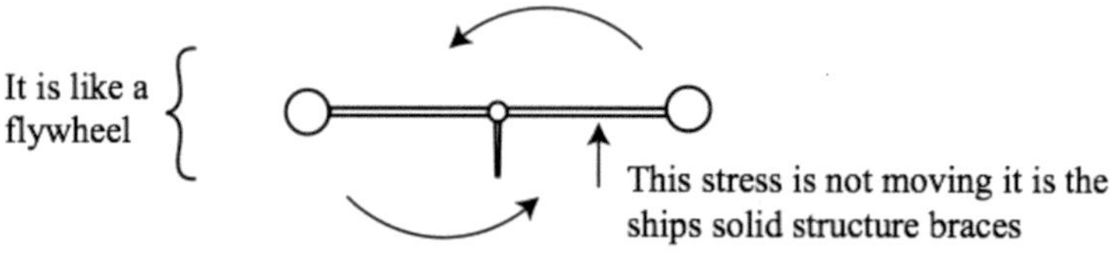

Boron glass is transparent to high speed protons that pass through it. Boron glass stops gas molecules from entering into the cyclotron. This means high speed proton collisions substitutes for chemical burn in the gas collision chamber where the plasma exits like a rocket nozzle. Think of the cyclotron as a big flywheel with the protons spinning very rapidly but released gradually through the boron glass. The electron gun is a simple hot rod that maintains ship neutrality by letting electrons escape to balance escaping protons. So what? No neutrons. It's not radioactive at all. In contrast the fission nuclear rocket is highly radioactive when operating.

Cyclotron Propulsion Device (Illustrative Not Detailed)

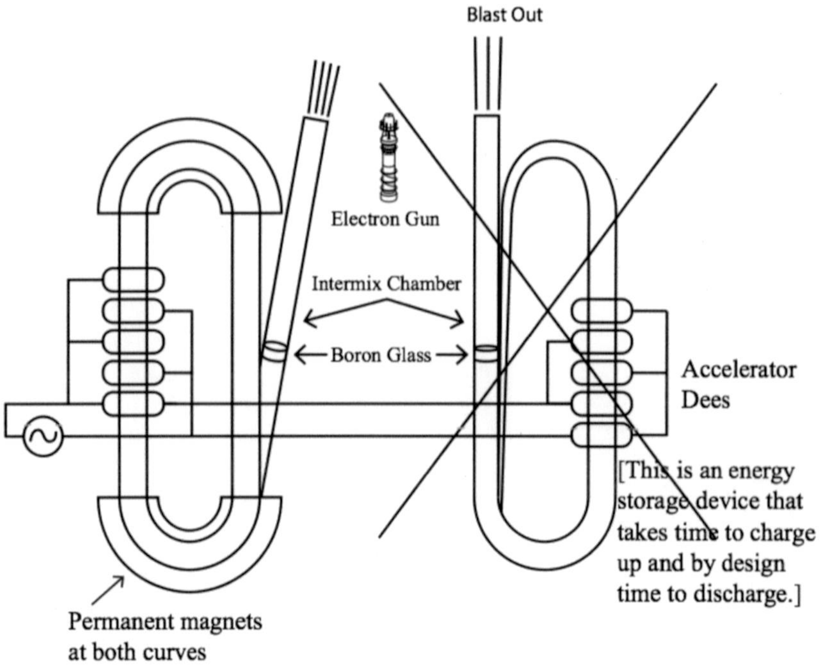

Correction: The two cyclotrons should be symmetrical like you are looking at this design and it's reflection in a mirror. Both require magnets. The thruster pipes are the rockets should be parallel. The cyclotrons can be angled to make the rocket pipes parallel in the drawing. If you want a side view it's a stack. Right now I am making a illustrative stack. This thing is capable of over 8 G acceleration or higher for a brief period. It won't work without an electron gun. Dr. Bronco said in 1970 the spaceship must maintain electrically neutral for charge equilibrium and an electron gun does that.

A grain of sand traveling at 90% light speed has the kinetic energy of an atomic bomb. Here protons are sent accelerating at that speed in a closed loop and are bled off gradually to collide with a propellant gas past the boron glass resulting two plasma exhaust jets. This takes an hour in which time the craft has lifted off from a planet big as earth and is well into escape velocity.

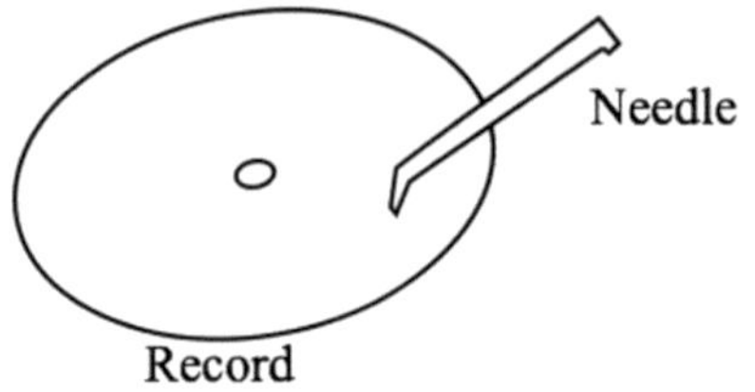

Like a needle following the grooves of a record fail safe designed not to allow the proton accumulation to exit all at once.

The UFO Motor Storage Device Explanation (Not Detailed)

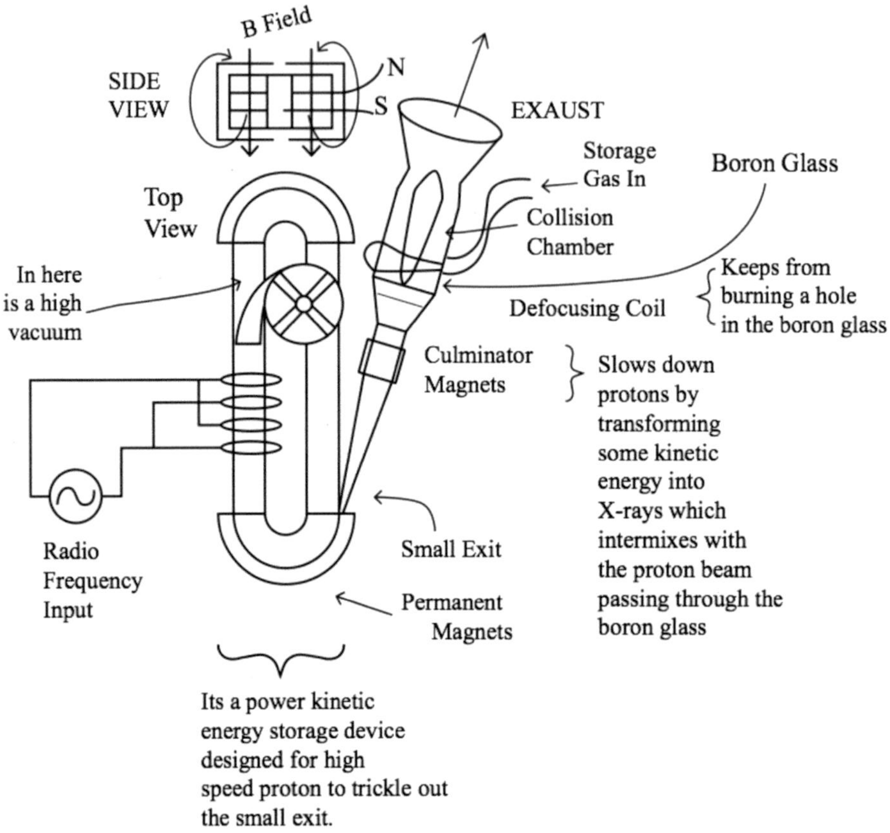

How the boron glass works: The fewer protons an element has the bigger by volume its size. The boron glass atoms are large but effective in stopping gasses from passing through it. High speed protons penetrate through it while at the same time stopping electrons.

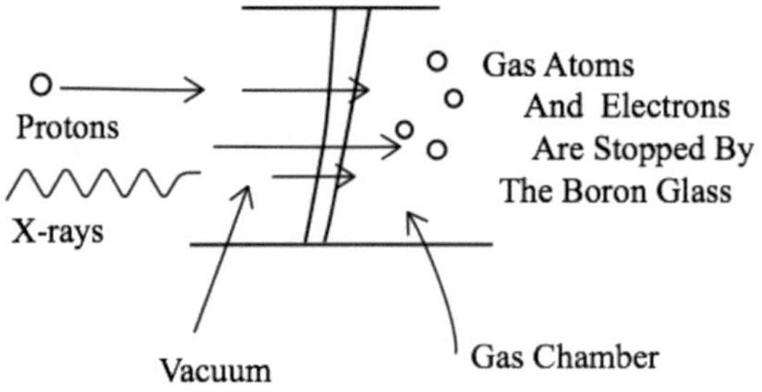

Why Two? So The Spaceship Can Turn Left/Right With 4 Full Maneuver

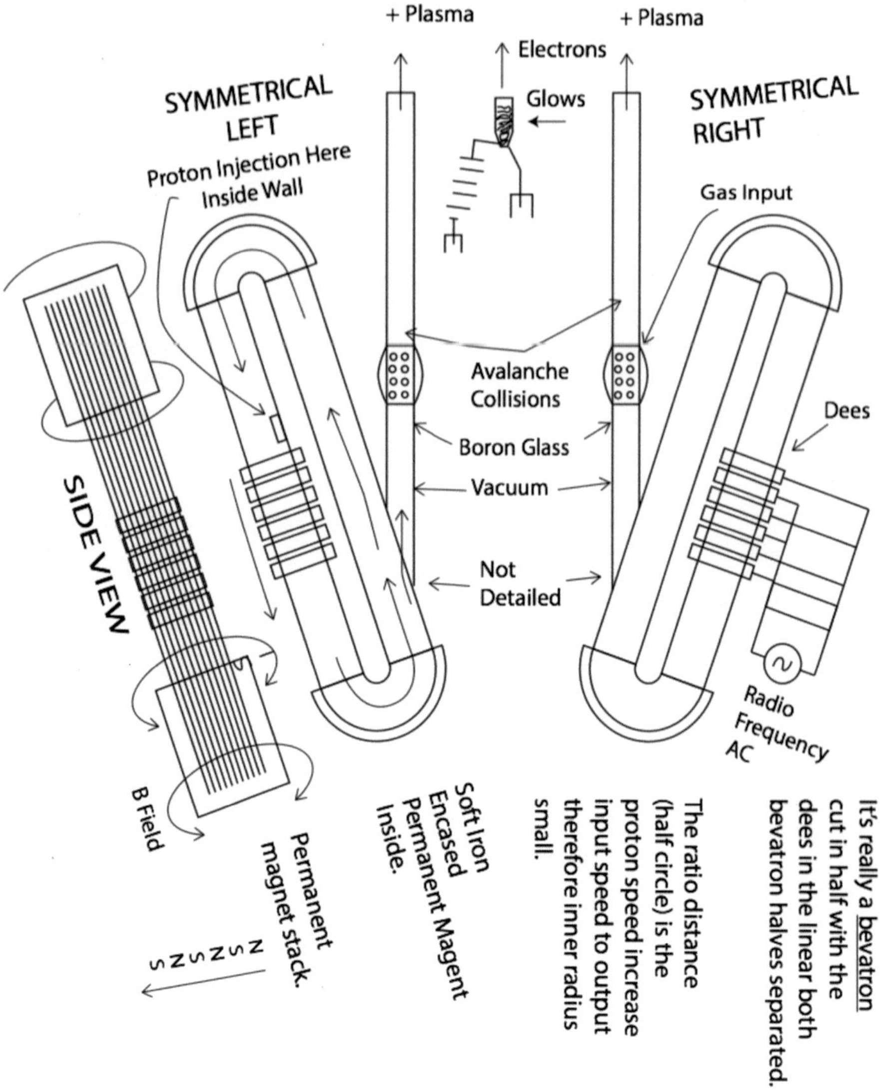

Toroid Bevatron

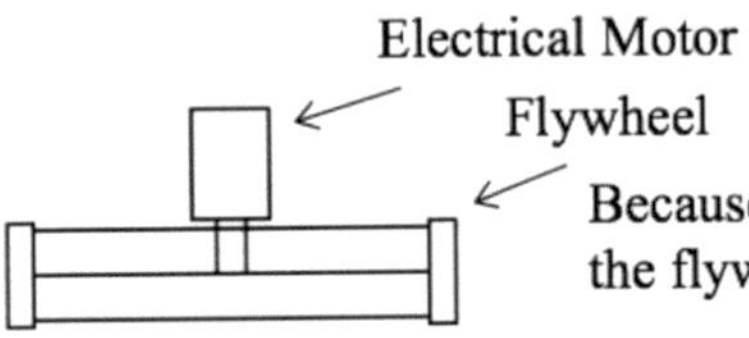

Because $F = MV^R/R$ (Centrifugal Force) the flywheel explodes

Centrifugal force explosion problem eliminated

In a bevatron toroid the protons spin in a void around in the donut magnetic field. The mass is very small and V is about 90% light speed. The centrifugal force problem eliminated it then is necessary the protons get bled off not sent out at once to collide with propellant mass (can be steam) because the interstellar spaceship has an abundance of water from low gravity worlds, comets etc.

Fools That Attempt This Should Be FIRED

[It's stupid to accelerate protons with this device to atom smashing velocities.]

[It's stupid to overcharge this device]

Humans are stupid and like asses try to do the wrong things because the military wants weapons. This was not intended to be a weapon but a rocket motor. That's why I was given permission to reveal only a <u>primitive</u> donut that one can get some useful work from.

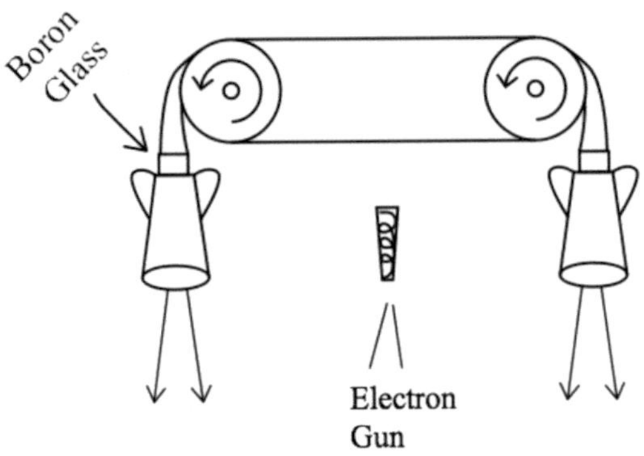

The bevatron existed in the 1950's. It's old hat.

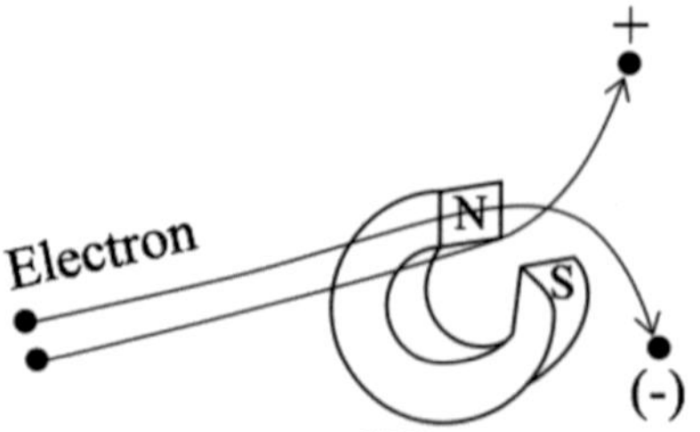

What
a magnetic
field does

The charge of protons
and electrons are
equal in magnitude
+Ω1.60E $^{-19}$ coulomb → proton
- Ω1.60E $^{-19}$ coulomb → proton
opposite in sign (∓)
electron mass is
1/15 36 of proton mass

Proton rest mass =
1.673E $^{-27}$ kilogram

Electron rest mass =
9.109E $^{-31}$ kilogram

For an electon
in a vacuum
velocity across 100 volts

$$V = \sqrt{\frac{2eV}{m}} = \sqrt{\frac{2(1.6E^{-19}\text{ coulomb})(100\text{ VOLTS})}{9.1xE^{-31}\text{ kg (electron mass)}}} =$$

5 million
9 hundred
thousand
meters/sec

For proton speed substitute
proton mass here

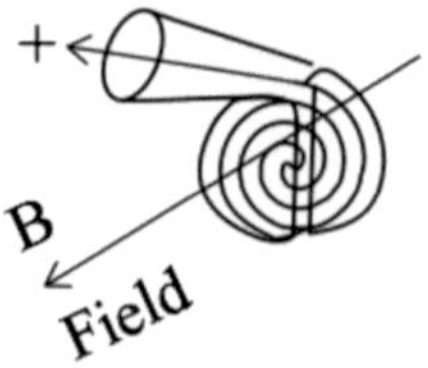

Many ∿∿∿ polarity reversals

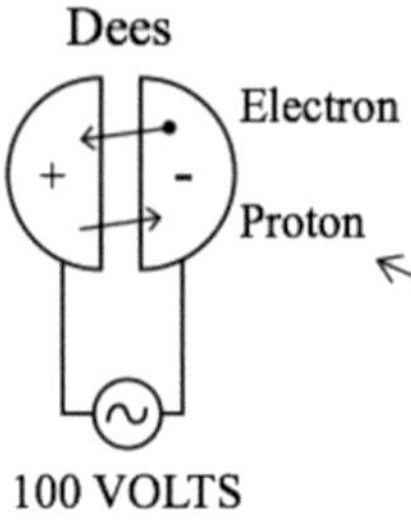

Dees in a vacuum
This is how the
dees work

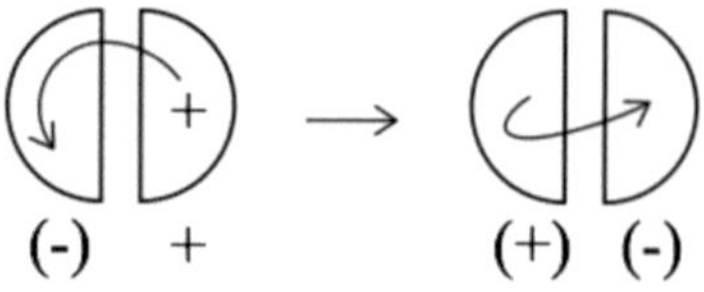

Because its high frequency AC
the polarity reverses the magnetic
field acts like a diode giving a
prefered direction

What's new about the toroid bevatron is weight reduction for use as a propulsive device - not as an atom smasher. IT SHOULD NOT BE USED TO SMASH ATOMS. Therefore keep proton exit speeds below smashing speeds.

If a toroid bevatron can store 1 gram of protons at 1 million volt potential - The static thrust in slow release is 1,910 kilograms for 1 hour duration.

Much higher if it is used to collide with a gas mass in the collision chamber.

Velocity In Meters/Sec =

$$\sqrt{\frac{2(1.6E^{-19}\text{coulomb})(100\text{ VOLTS})}{9.1x10^{-31}\text{ kilogram}}} = 5.9 \text{ X } 10 \text{ Meters/Sec}$$

One Electron Weighs 9.1 X 10 $^{-31}$ Kilograms

One Proton Weighs 1.673 X 10 $^{-27}$ Kilograms

Proton Velocity In Meters/Sec =

$$\sqrt{\frac{2(1.6E^{-19}\text{coulomb})(100\text{ VOLTS})}{1.673 \text{ X } 10^{-27}\text{ Kilograms}}} =$$

$$= \sqrt{191.273 \text{ X } 10^{27-19}} = 13.83 \text{ X } 10^{4} \text{ Meters/Sec}$$

Proton At 1,000,000 Volts
Velocity In Meters/Sec =

$$\sqrt{1.91273 \text{ X } 10^{27-19}(1XE^{-6}\text{ VOLTS})}$$

$$\sqrt{1.91273 \text{ X } 10^{E14}} = .38 \text{ X } 10^{7} = 1.38 \text{ X } 10^{7} \text{ Meters/Sec}$$

1 gram = .001 Kilograms
= 1XE $^{-3}$ Kilograms

Kinetic Energy For 1 Gram Of Protons Stored in The Toroid Bevatron = 1/2 (1XE ^{3} kg)(1.38E ^{7})^{2}
= (.5)(1.38 X 10^{14}) = .69E14 joules

Static Thrust = $\frac{.69X10^{7}\text{ Kilograms}}{3600}$ = 1.91X10$^{(7-4)}$

(1, 910 Kilograms Thrust Sustained For 1 Hour Duration)

Propulsion Motor Matrix

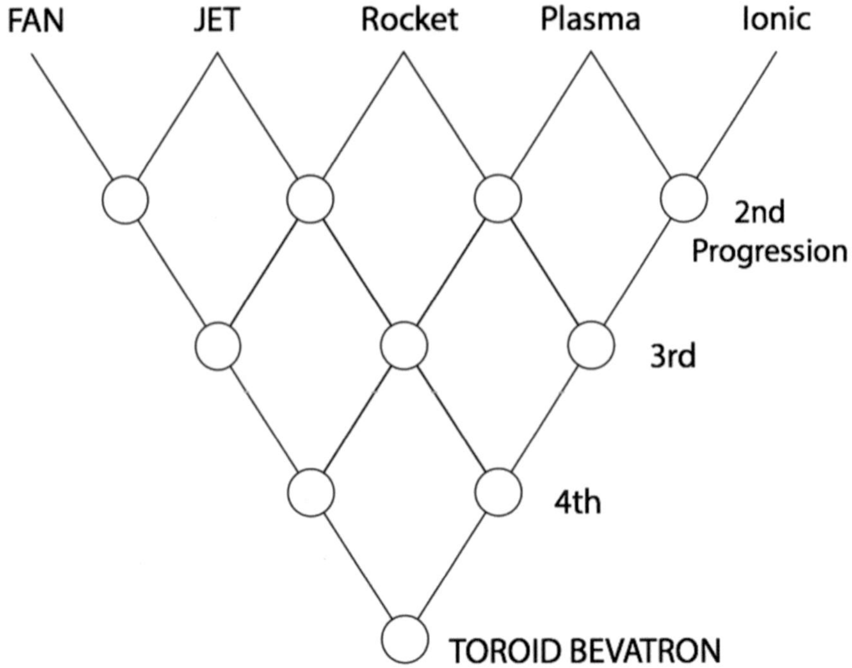

TOROID BEVATRON
Both stores energy and control release at near atom smashing exhaust velocities. (Exaust variable from subsonic to near atom smashing velocities)

I went on to improve the 1970's design alone without help from anyone

Each circle is a propulsion motor design I left in the hands of the ex mayor of Grand Prairie, Texas in the 1970's.

The Propulsion Matrix Is Three Dimensional

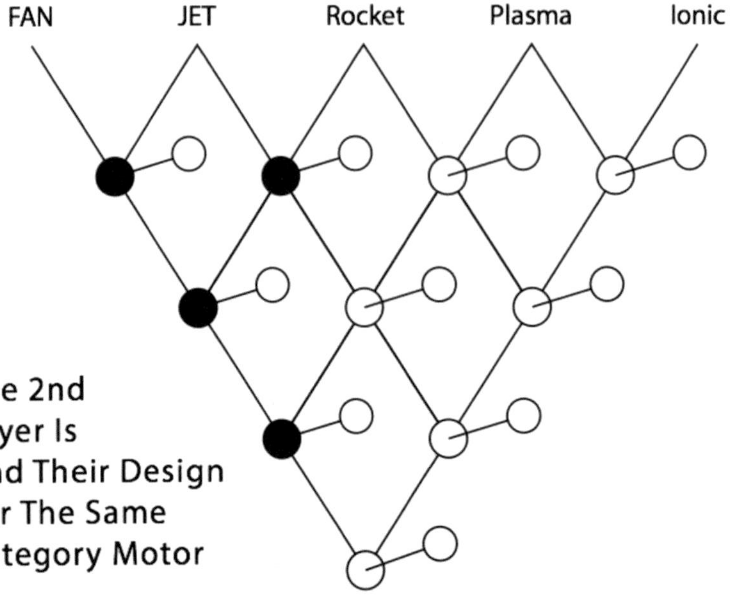

The 2nd
Layer Is
And Their Design
For The Same
Category Motor

<u>Fan-Jet</u>
Is a turbofan with gear reduction to a gas turbine as seen on a typical 737, 717 etc.

WYLIE TEXAS 1987

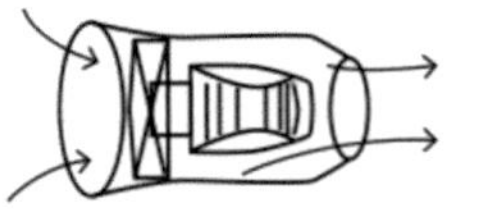

Without gear reduction and with afterburn - Supersonic combat jet.

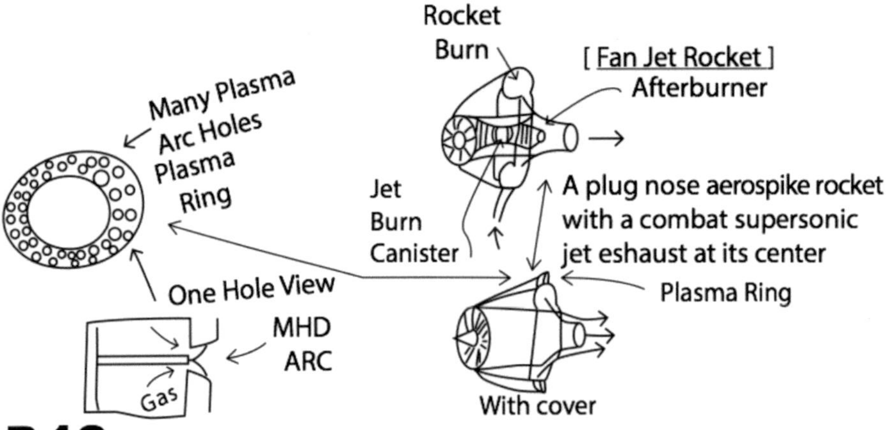

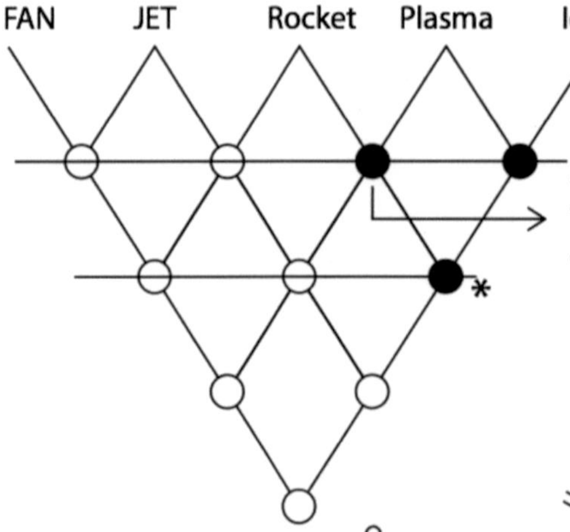

<u>1987 Wylie, Texas</u>

<u>2nd Progression</u>

The Same Thing Without High Voltage Applied (No Ionic)

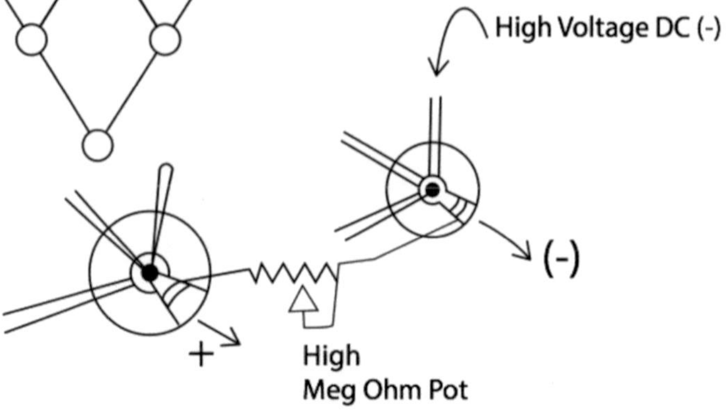

[The Plasma Ionic Motor]

Hot ball glows by resistance heat (isolation transformer)

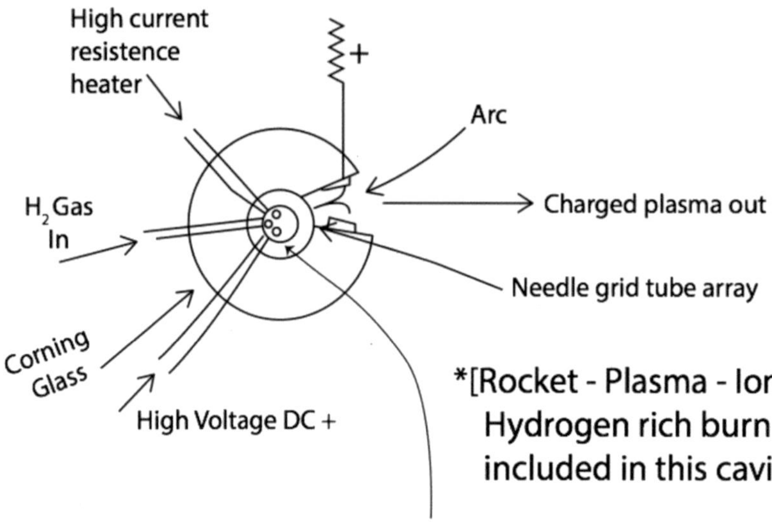

*[Rocket - Plasma - Ionic]
Hydrogen rich burn
included in this cavity

The [Jet - Rocket - Plasma] - Ionic Motor
and
The Jet - Rocket - Plasma - Ionic Motor

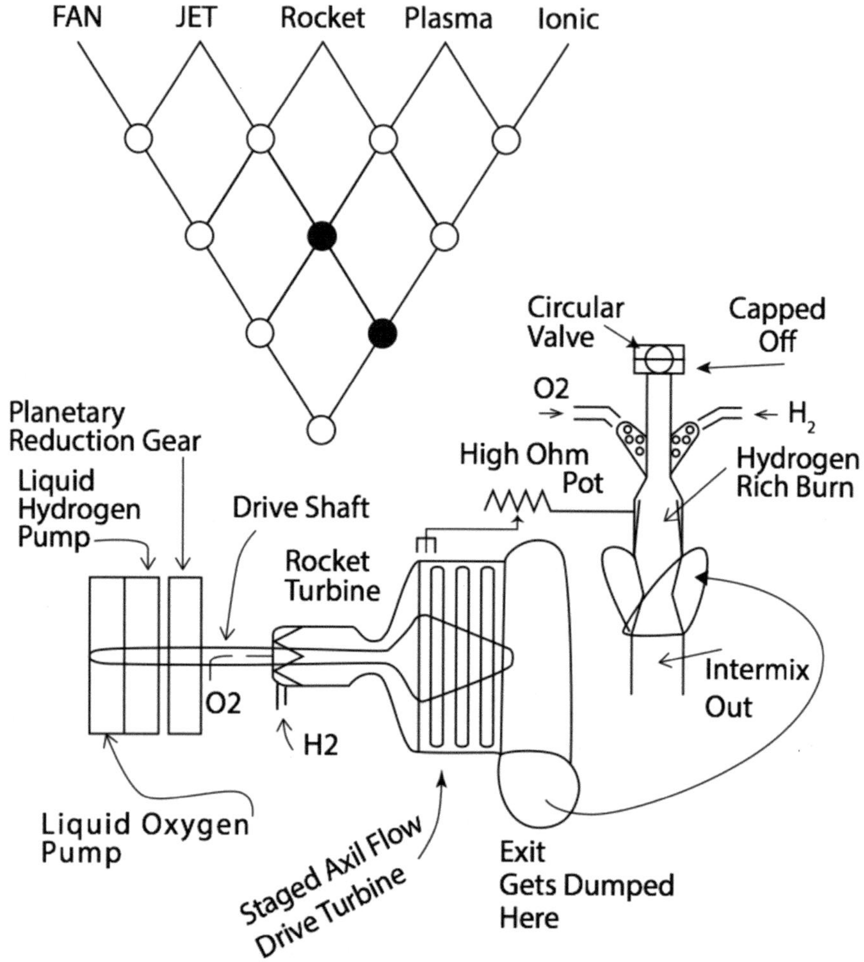

[1987 Wylie, Texas] storm shelter designs destroyed and memory recovered in 2013 (Dec) It's actually the intermix chamber that attraches to the boron glass proton bevatron. Capped off to illustrate pure jet - rocket - plasma function circular valve on to permit proton avalanche collision (plasma) results in jet - rocket - plasma function. Pot adjust results in plasma - ionic function. This completes all the propulsion motor sample illustrations. The parallel (third dimesion) is for quality upgrades or design variety as an example the staged pipe cluster burn design incorporated as a design variant.

UFO MOTOR

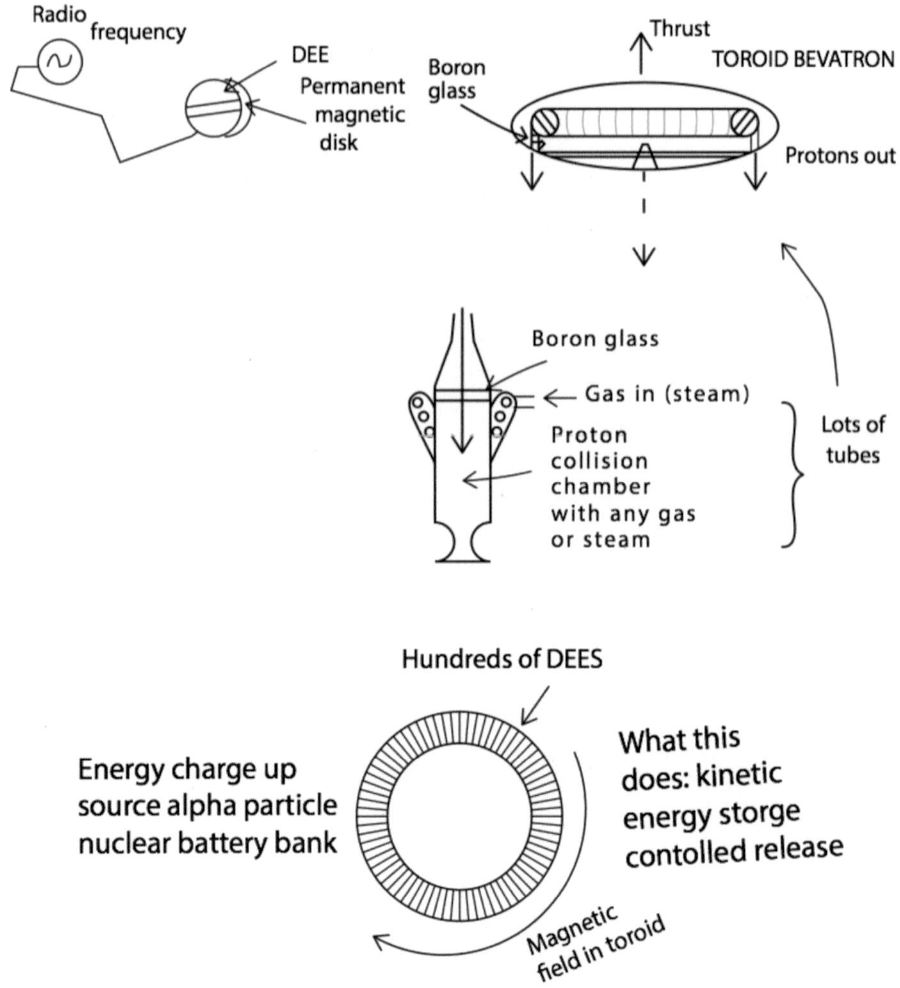

Once in space, the toroidal does not need the boron glass and the collision chamber. Once in space, the dees are capable of accelerating entire atoms ionized. To deal with this, very simple. Some of the dee exits are designed to operate in space and some with boron glass. So then operating in the atmosphere the sections without boron glass are shut off so air can't get in. The inside of the toroid is designed so gas can't leak to separate sections. The technology for generation of positive ions is quite advanced and is the basis of microchip electronic industry.

Primitive Toroidal (Donut)
Proton Particle Accelerator Propulsion Motor

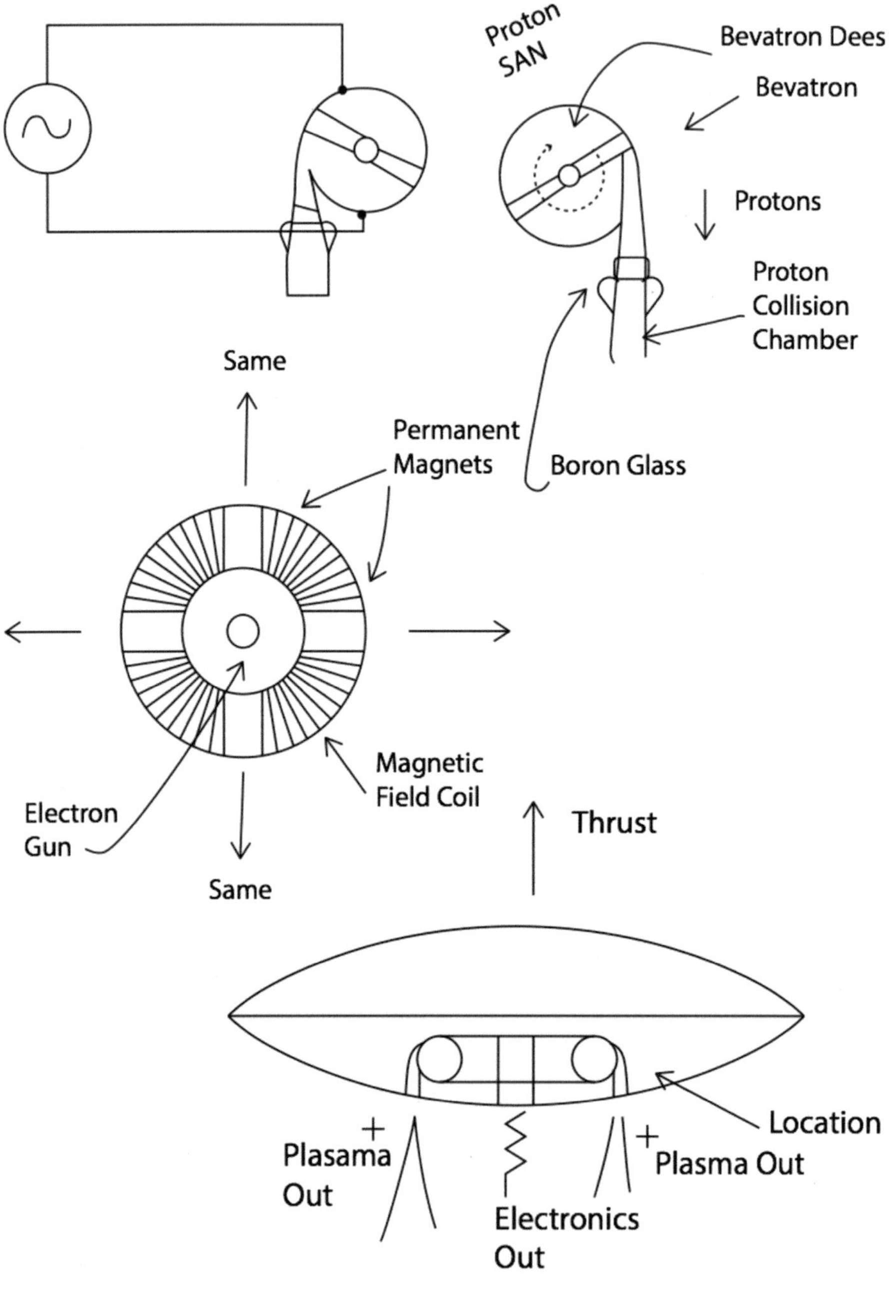

P.22

The Interstellar Spaceship Fleet Motor

The california nuclear fusion device is stupid because helium 3 is extremely rare.

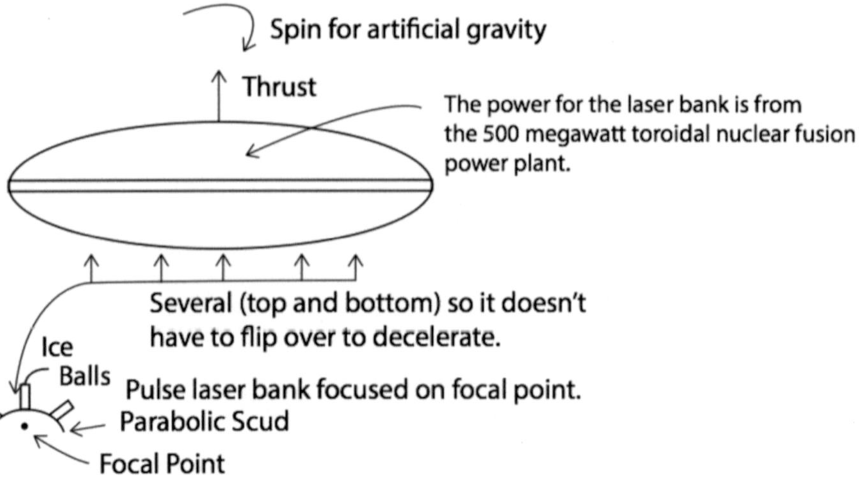

For the reason the interstellar spaceship has a large deuterium reserve for its toroid nuclear fusion power generator, ordinary ice balls are blastedat the focal points for thrust.

(Ice that was pocessed to have its deuterium content removed) ordinary ice balls are shot at the focal point like a machine gun. Sensors detect the ice ball and fire the laser banks the ice ball instantly explodes into a plasma ball.

The thrust is a function of the temperature rise.

$$V = \sqrt{\frac{3KT}{m}}$$ In Meters/Sec

Velocity Of Blast Gas = V
Boltzmann's Constant 1.38 X 10-23 joules/degree kelvin = T
m = Molecular weight of one molecule of H2O

[No attempt to cause ice balls to reach nuclear fusion temperatures]
It's not a nuclear fusion blast.

Because this kind of motor has a ISP far superior to chemical burn, interstellar travel speeds is possible

I thought you might include this as I forgot to describe the interstellar spaceship fleet motor. The shuttle craft motors are different depending on application. The fleet motor generates much less than one G so as not to disrupt internal activities where centrifugal force is used for artificial gravity.

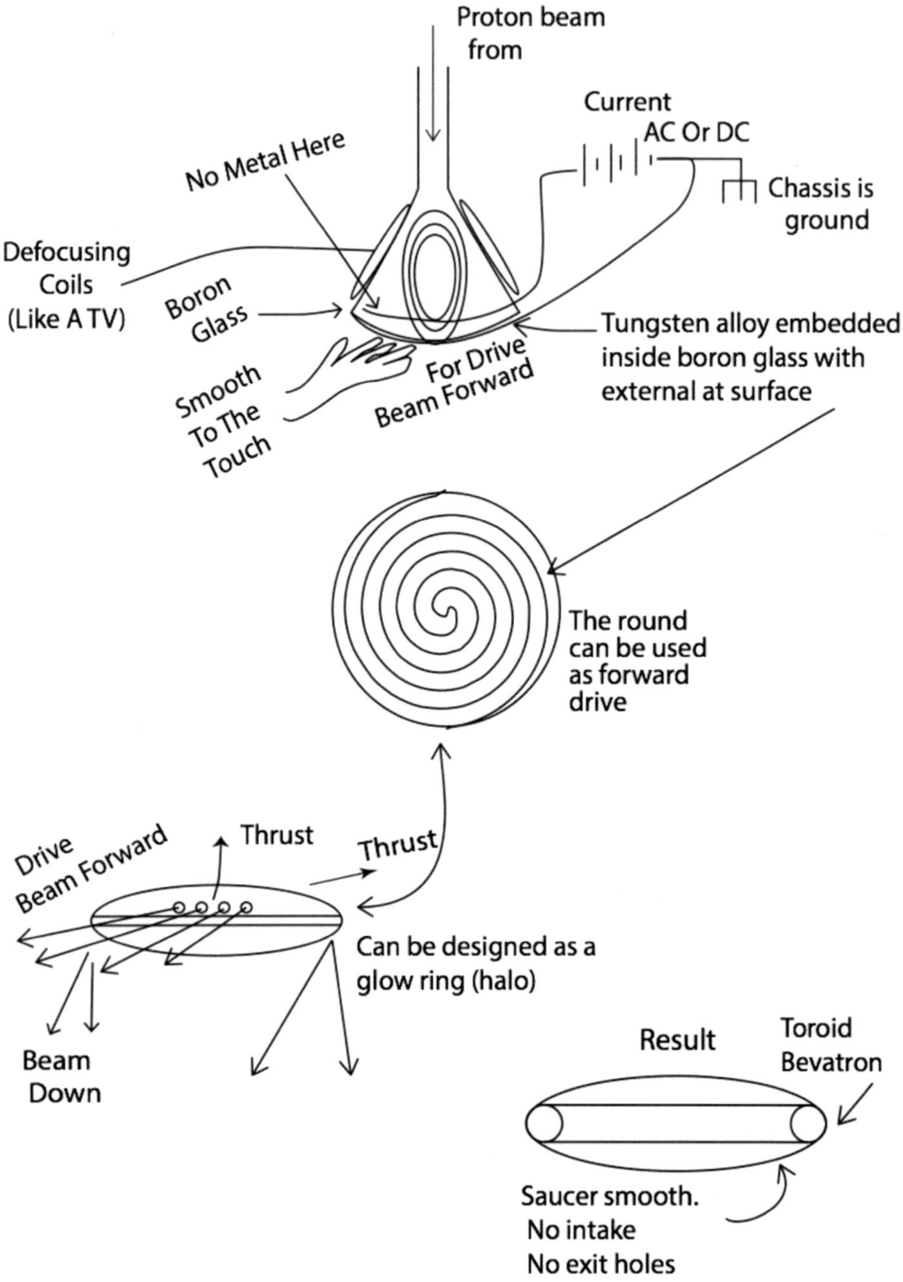

What's it for? The spiral is the negative electron gun that neutralizes protons that exit (transforms proton beam into plasma external to the saucer).

This Compares The Chemical Burn Rocket To Its Ice Ball Upgrade.

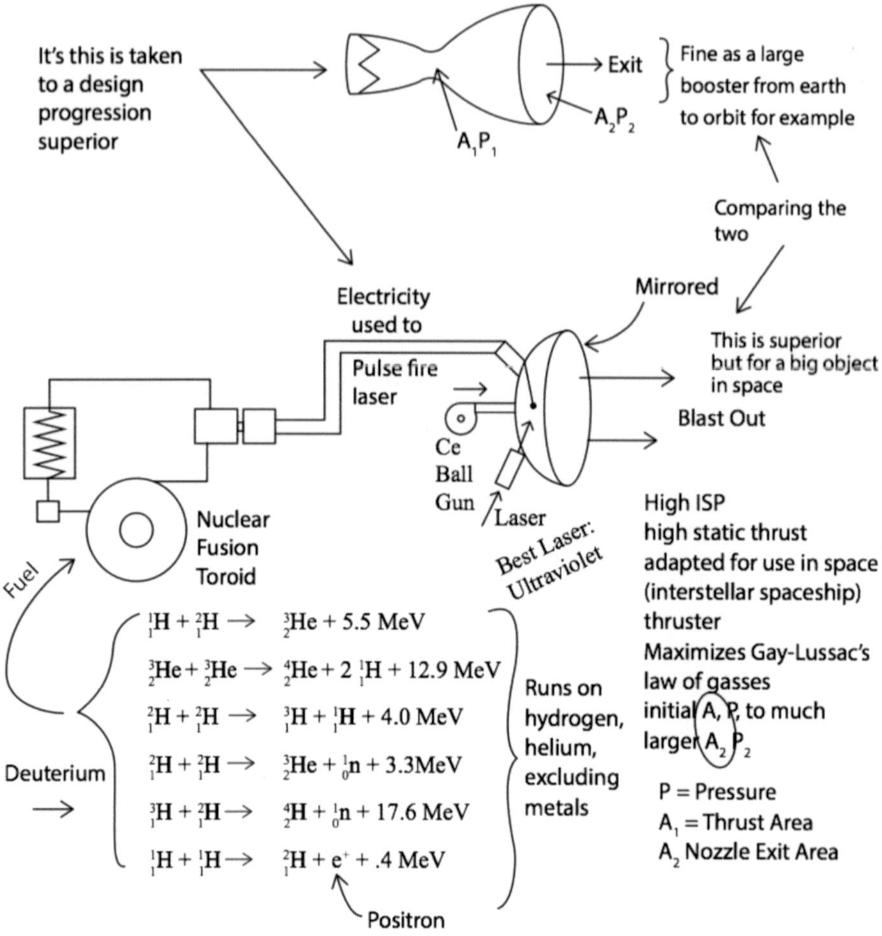

* Lithium fusion excluded for security reasons
(It's a metal that messes up the interior and damages the toroid)

In A Dream State I Saw The Following And Then Woke Up.
In Another Dream I Got To Ride The Saucer Like That.

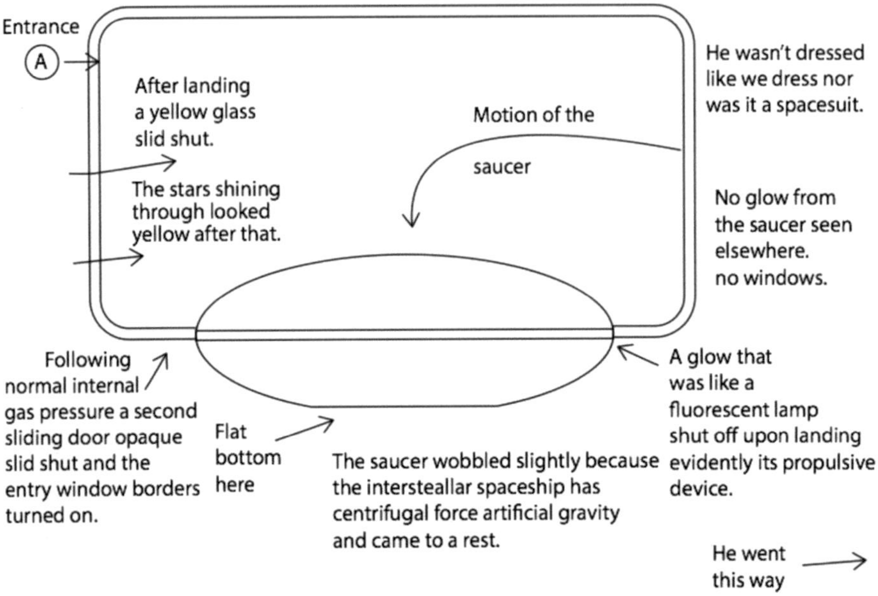

No landing pods. The alien exited the back side - I didn't see what kind of exit the saucer had. The alien walked to the left; pulled down a lever twice and let go. A sliding door opened and he went in a long corridor with the door closing behind him. The big saucer was so enormous there was no noticeable curve to the floor of the long corridor he wint into.

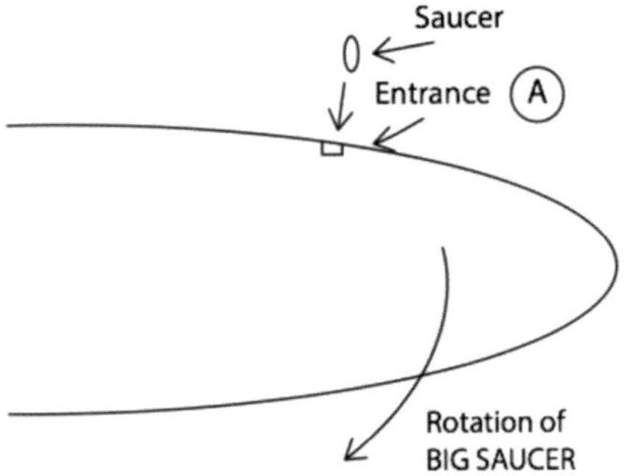

What I saw fits in perfectly with a saucer entering an interstellar spaceship that slowly rotates.

Cigar UFO

In the cigar UFO the toroid bevatron is long. The curved part is too close the magnetic field loop and contains no dees. The straight part contains dees both sides for vertical lift. The nuclear fusion toroid vastly reduced in size supplies height and electrical power. For long missions the cigar spins for internal artificial gravity.

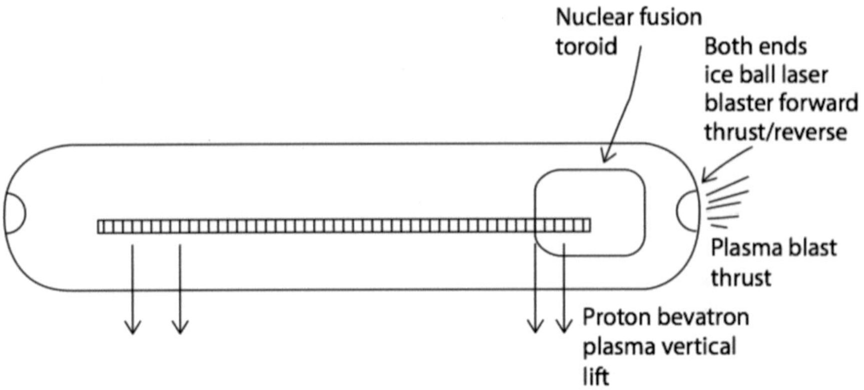

What is it for? It is sent by the interstellar fleet to low gravity celestial bodies to take in ice so the whole fleet doesn't need to come to a stop for re-fuel. Then it catches up and docks with the fleet. For this reason it requires its own nuclear fusion toroid.

[I did not reveal how the fusion toroid was modified so it doesn't take up the volume of a four story cube but showed the profile so it explains how the fleet can interstellar travel without delay, slowdowns, or stops.]

So you wont confuse how the dees are in the cigar ufo for lift; I chose to detail further.

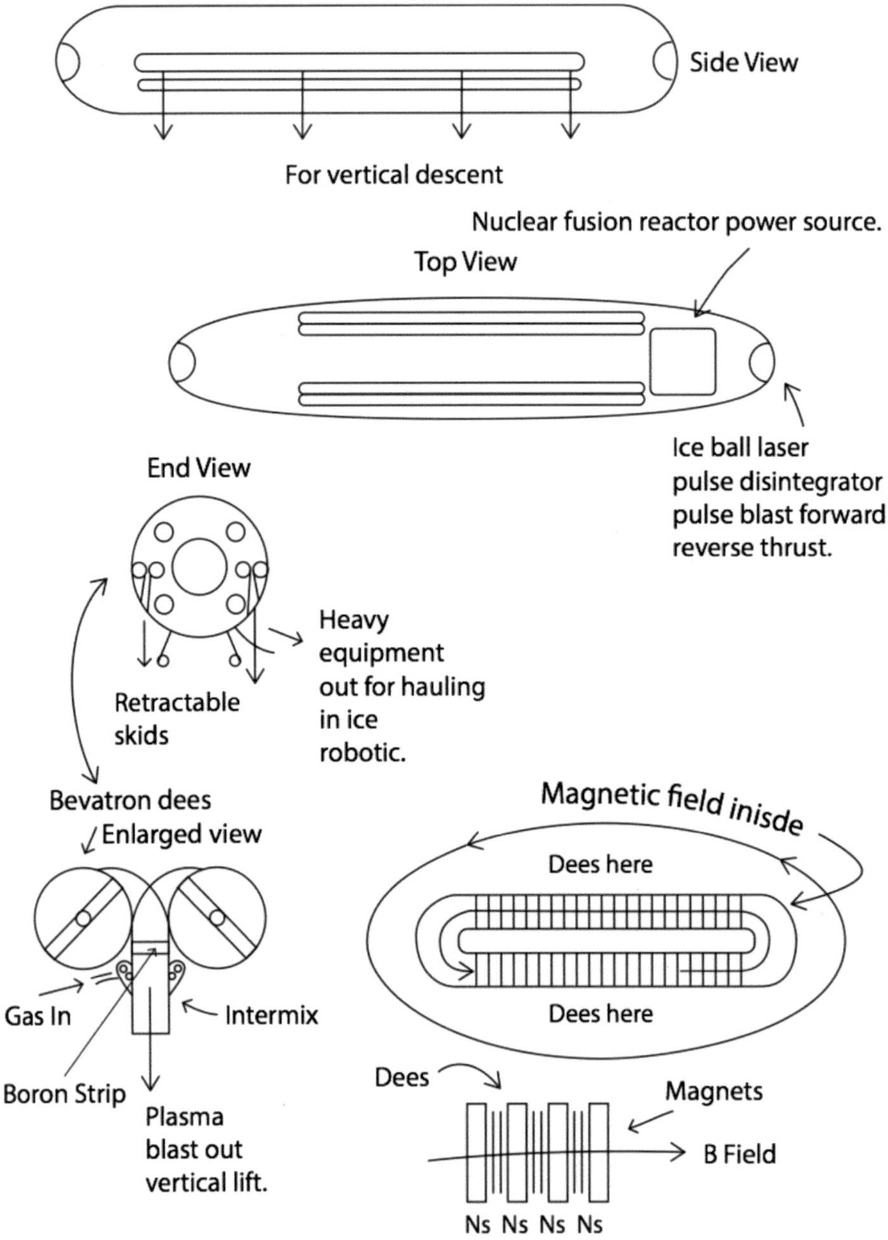

The Nuclear Fusion Toroid Reactor

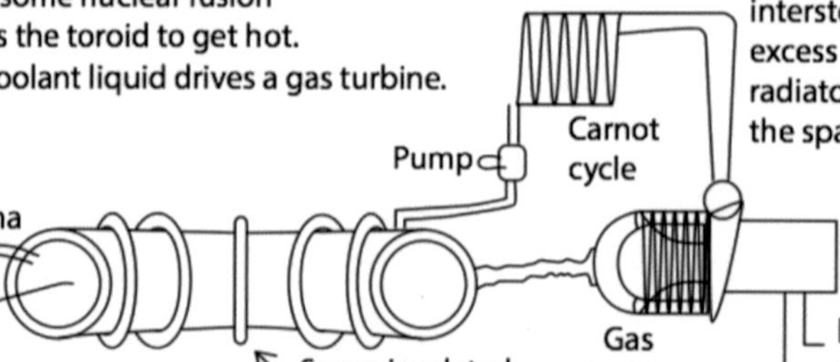

Why plasma?
A magnetic field repels a plasma here toward the toroid center.

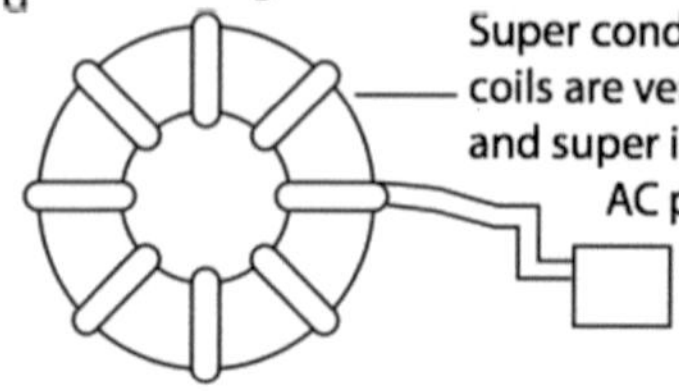

The deuterium in one bucket of water has the energy equivalent of several boxcars full of petroleum (from gravity KY worlds)

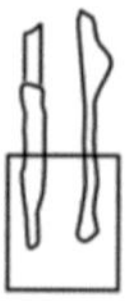

That energy is recycled by fuel cells (gets most of if back) by electrolysis hydrogen is separated from oxygen in the H_2O (water)

The hydrogen is liquified and about 2% by slow boil off remaining is deuterium tritium mix

The hydrogen deuterium free goes to fuel cells to get back most of the energy used in electrolysis

Is also used by high ISP rockets for mobility and for shuttle crafts

The amount of deuterium energy content in an ordinary bucket of earth water exceeds the energy content of 20 railroad petroleum fully loaded boxcars made for hauling petroleum.

Primitive UFO Motor

2 years ago
I built and tested a primitive UFO motor

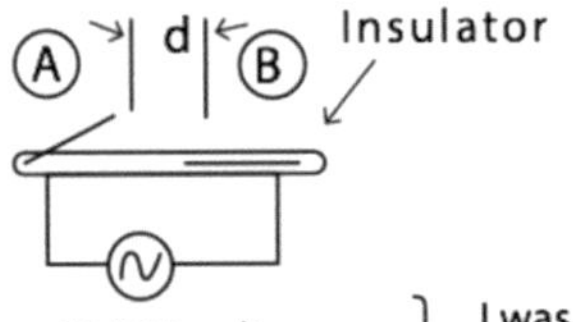

Waves of + and - ions collide with atmospheric gasses resulting in a wind originating at (A) toward (B)

10,000 volts
AC sine wave

} I was limited to my 10,000 V AC power supply

(B) is totally insulated causing the wind to deflect.
(A) is needle points whose tips protrude out about 1 millimeter. The separation distance needle point to leading insulated B plate is 1/2 cm.

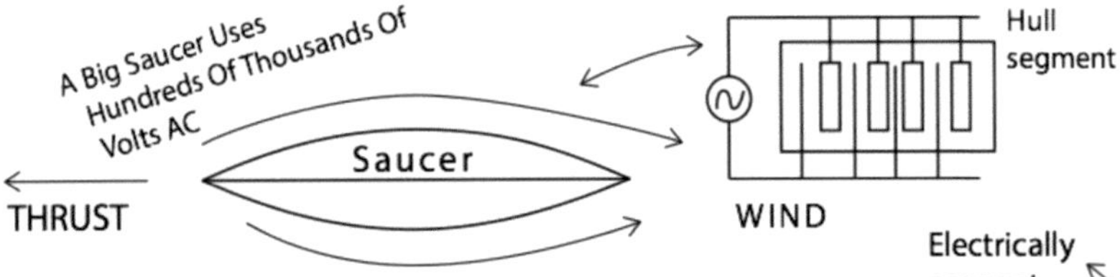

Electrically connect

I built and tested the cascade arrangement.

In the place of needle points are plasma motors so then the ship generates a sustained thrust in both an atmospheric environment and in a vacuum.

The nuclear battery is used to charge up battery banks for power surge. The plasma motors can also burn H2 wuth O2 chemically in their burn canisters.

In the NASA nuclear battery the plates are alternately coated with alpha particle element (radioelement isotope).

In the saucer nuclear battery the plates are alternately pn (diode) sandwich permitting the battery to sustain a much higher emu back voltage pressure hence more powerful for its size.

Alpha particle plates retractable

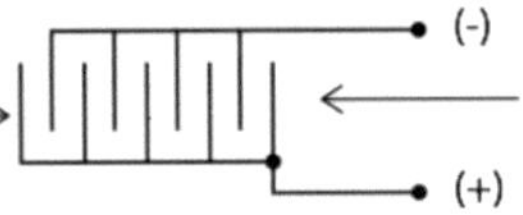

PN semiconductor coated plates

The UFO motor is not far fetched sceince fiction but a logical progression in propulsion design.

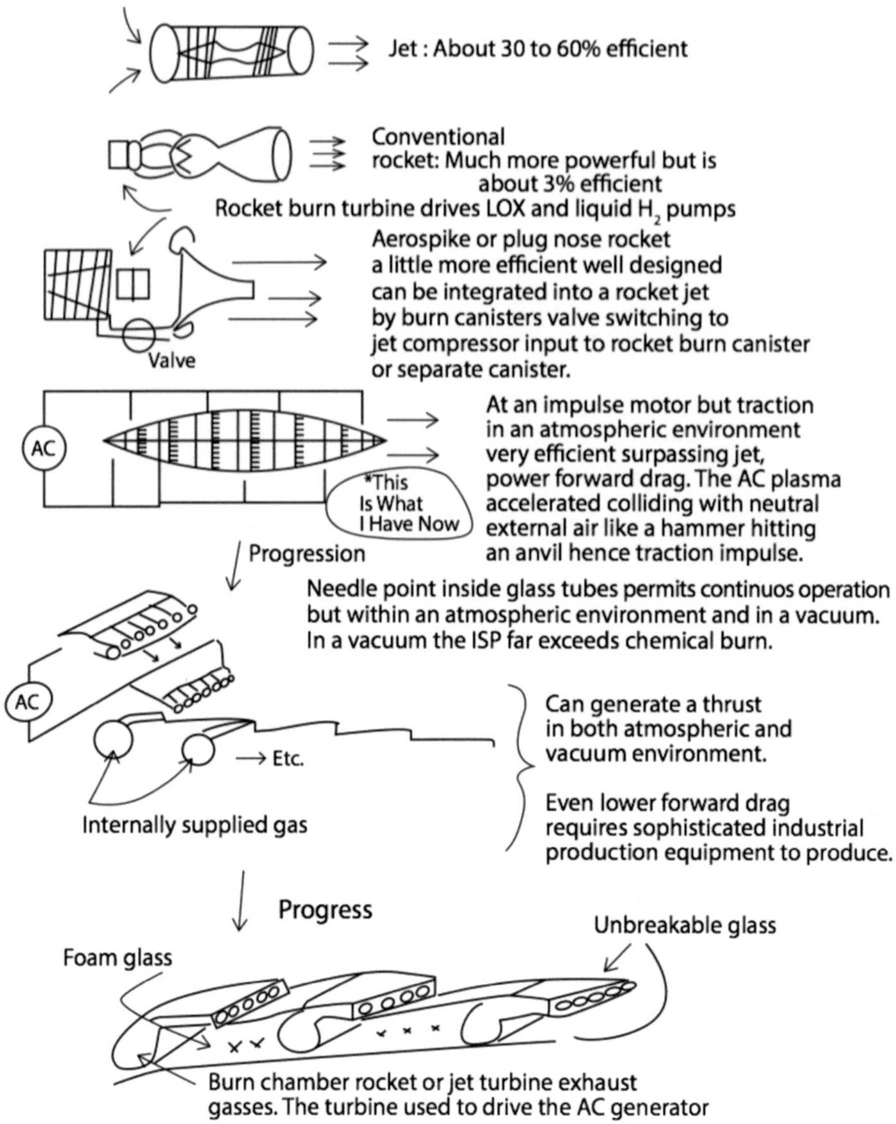

What I had in the tornado shelter was a DC ionic breeze coming from one single needle point I could feel with my hand over a yard away; and I could make the ions positive or negative by flipping over the high voltage diode. If Wylie, Texas had left me finsih my saucer I would have outpaced the USAF jets of that time. My AC ionic motor compacts the design. With DC requires separate motors at considerable distance seperation. AC eliminates that. Now then as for the beer bottle apparatus (Page 43); the beer bottles are high voltage capacitors for the reason home built - electronic stores wont sell stuff reserved for business and research institues that have a lot of money and lawyers.

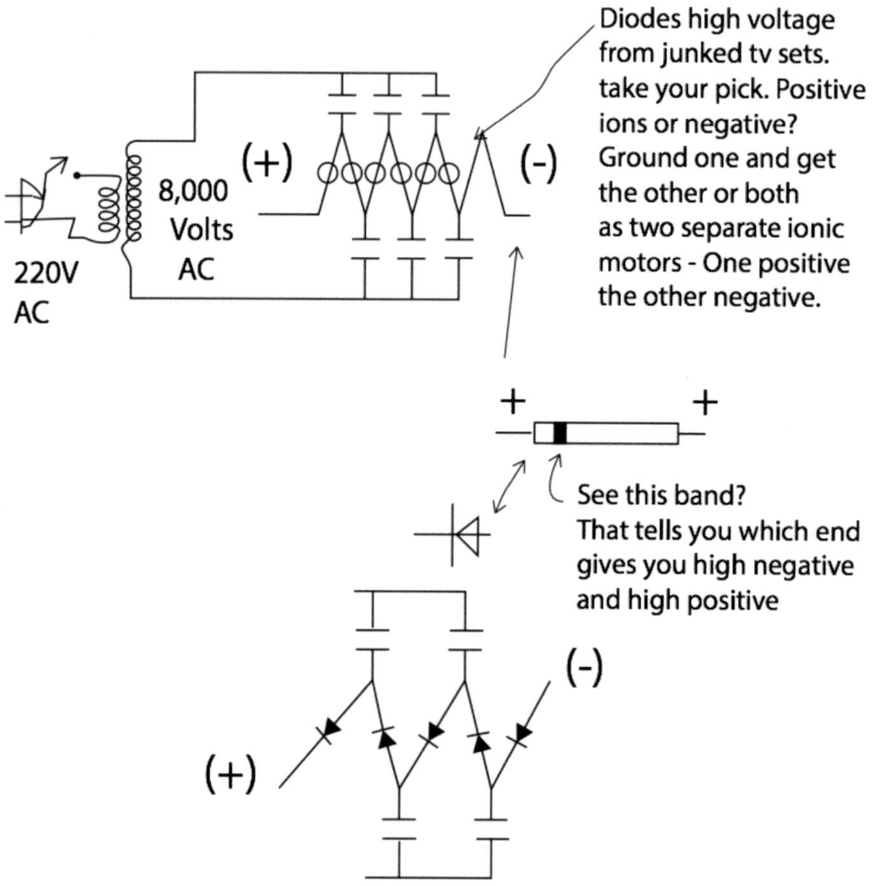

In the tornado shelter I was using a much higher frequency and all my stuff was from junk. My AC motor is weak because I'm using junk - what other have discarded for power.

In theory:

The flying saucer utilizing only LOX and liquid H_2 can attain escape velocity from a verticle liftoff no need to improve nuclear energy the reason being rocket burn is 3% efficient but the same rocket burn as a gas turbine driver can be up to 80% efficient conversion to torque. If that torgue drives a planetary gear counterrotating fan for ground liftoff then the AC ion motors in forward drive result in aerodynamic lift making possible clutch attachment from the verticle counterrotating fan diverting all the torgue power to the AC high voltage generator that drives the ac ionic motors. The ionic motors is not Gay-Lussac but ion waves + followed by - based on Faraday's law where changing intense electrical field not thermal expansion is the accelerating force.

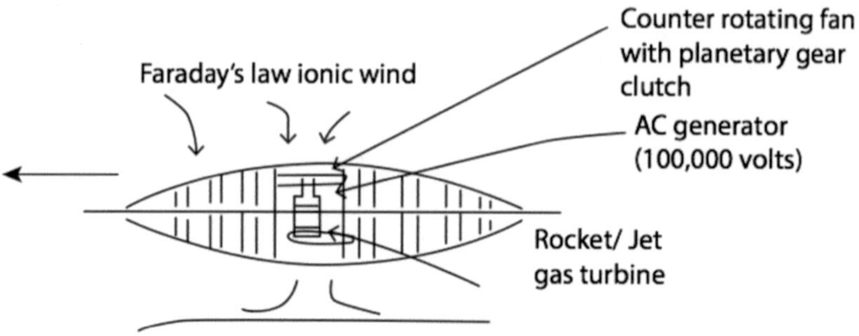

For lunar/martian landing; initial liftoff aerospike/ plug nose can substitute or supplement as seen in the fantasy flying saucer model "Forbbiden Planet."

The faraday motor has two configurations; tubular and hull. I have tested both successfully

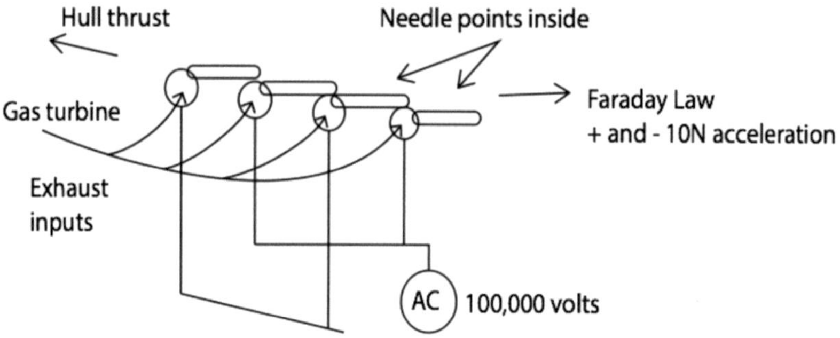

AC Thrust In Simple Terms

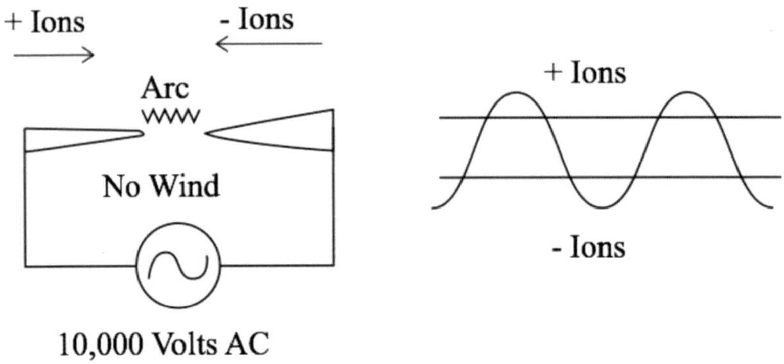

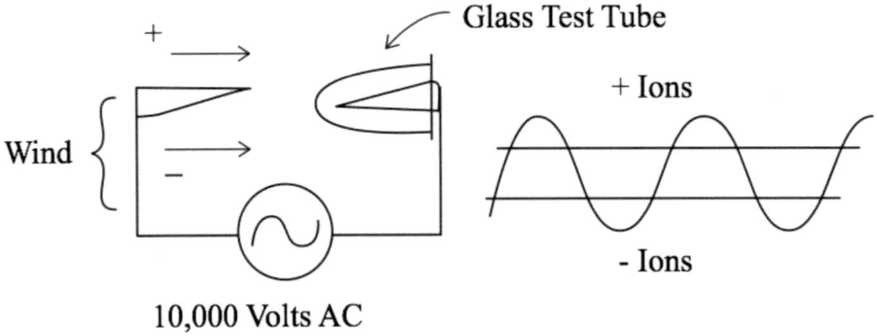

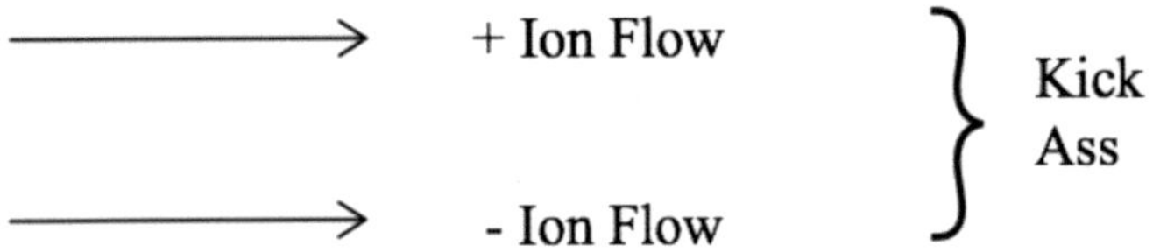

(Neutral air is hit hard like a hammer hitting an anvil)

The insulation where shown stops ions from forming there the thrust can only be one way [⟶] for both + and negative ions is that simple enough? AC thrust.

This I discovered by accident after I dreamed aliens in another galaxy have an electronic symbol that looks similar to the United States flag (by coincidence).

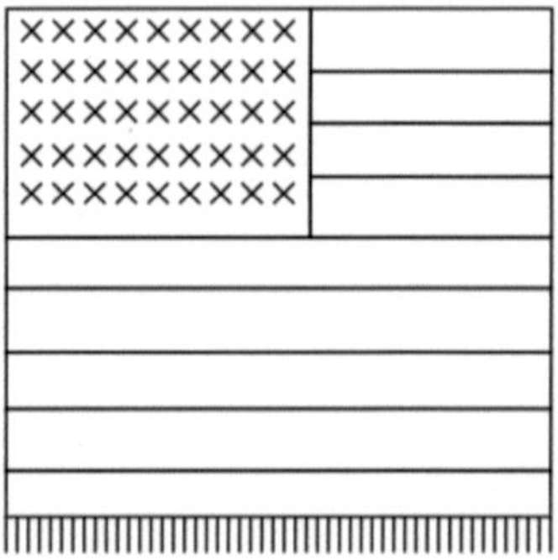

Stars: Stars are made of plasma. Plasma responds to electrical and magnetic field in the dark my needle points arraged to array create mix plasma points in the presence of an intense electric field.

Blue: Gas ambient
Red: AC insulated strips
White: Spacing between strips

← The US Marine flag has this to put the needle points in array (by coincidence).

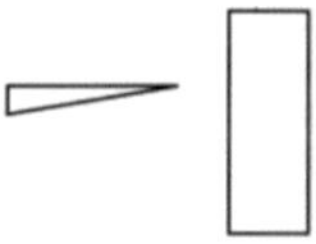

If noth strip and needle point is nude, then positive and negative flow in opposite directions resulting in an arc - no thrust.

If the needle point is nude and the metal strip is insulated - ions can only move in one direction. A thrust is generated.

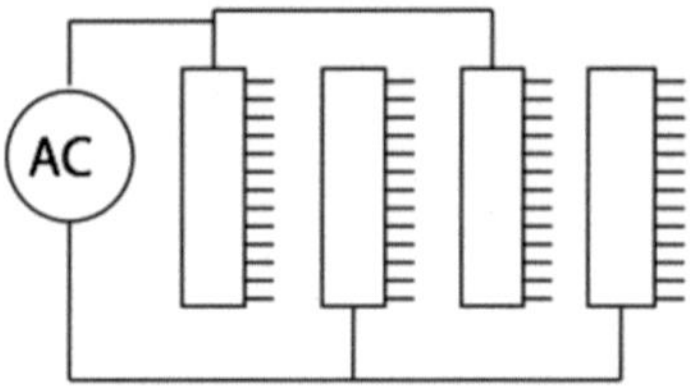

To increase thrust: Use a high puncture resistant material like unbreakable glass - I dont have that so I use plastic to demonstrate the motor works. With high puncture strength glass the voltage thrust can be increased; intensifying the flow.

If the frequency is increased, the needle points needed to protrude resulting in a smooth surface. If the needle points are in any holes a gas can be pumped in and the kind of gas results in a color or glow. That's why UFOs can change color. Why AC? It's easier to build lightning powerful AC generators eliminating a lot of headaches associated with DC thrust motors.

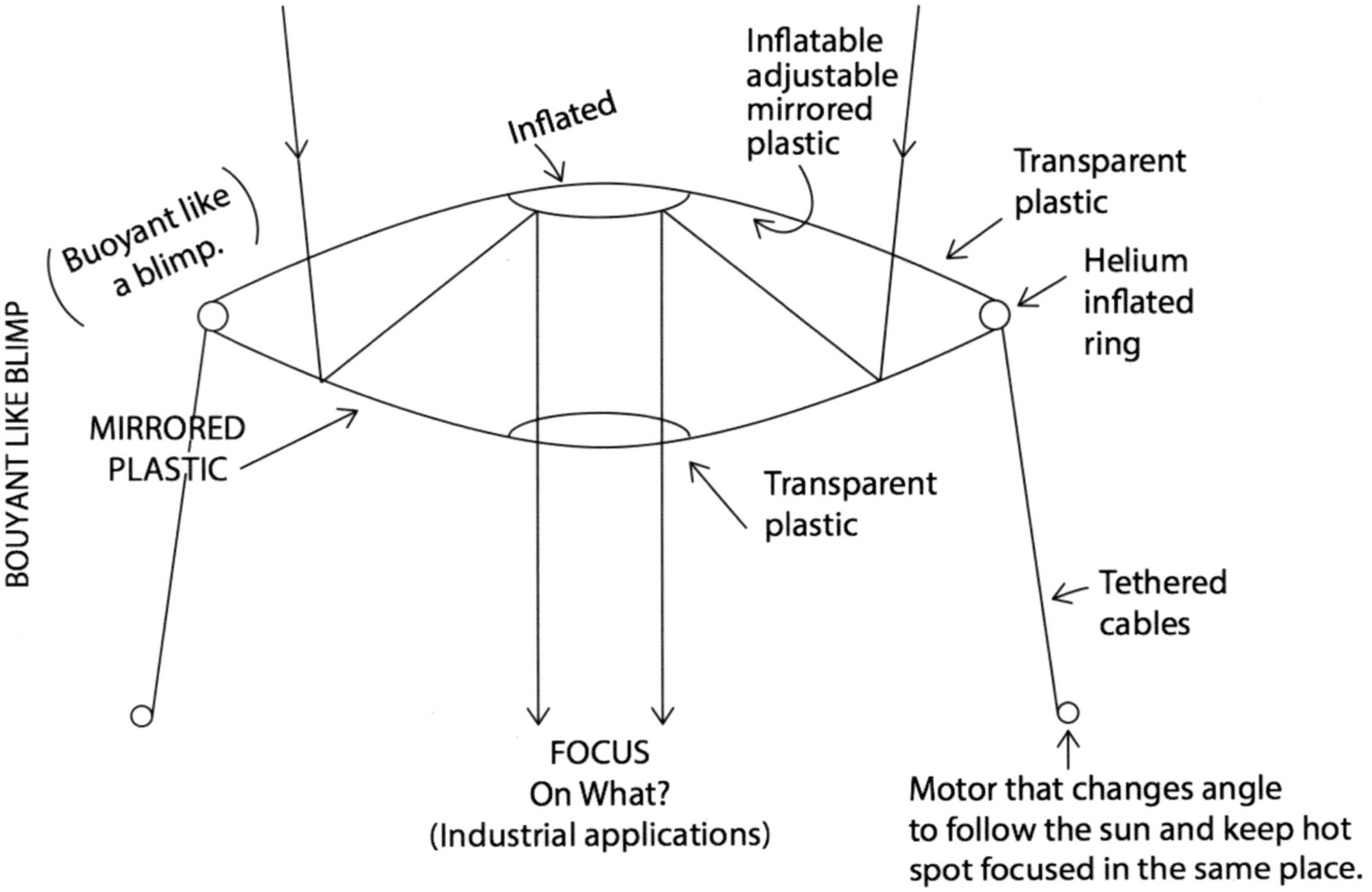
BOUYANT LIKE BLIMP
Buoyant like a blimp.
Inflated
Inflatable adjustable mirrored plastic
Transparent plastic
Helium inflated ring
MIRRORED PLASTIC
Transparent plastic
Tethered cables
FOCUS
On What?
(Industrial applications)
Motor that changes angle to follow the sun and keep hot spot focused in the same place.

P.36

Super Massive Black Hole At The Center Of The Material Universe

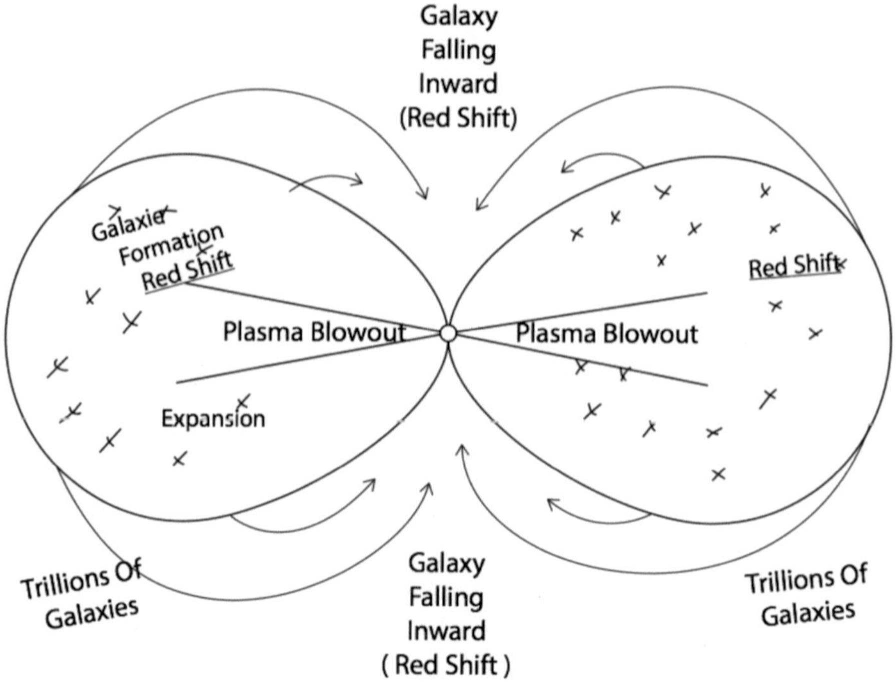

What falls in gets disintegrated and exits as a plasma blowout (inverse entropy) STATE OF THE UNIVERSE: Perpetual Steady State Kaleidoscope (ever changing - ever changeless overall)

The Big Bang Theory is false. Einstein intuitively believed in the steady state theory but did not know how to integrate the Hubble redshift. This integrates the Hubble red shift in Einsteins Steady State Universe Theory.

An American Astronaut on the moon dropped a stone and a feather. Both landed at the same time as in a similar experiment in a vacuumed chamber here on earth. The Cosmologists intent on pushing the Big Bang theory on the masses have been proven wrong but don't want to give up their theory. How did this experiment prove them wrong? Over vast distances the hubble red shift is caused by the fact that if one of the two objects was separated by an immense distance, wether the two objects were traveling upward or downward their respective distance would change. Either way scientists would see what they call red shift.

Electric Field String

The universe can be compared to a hologram without being that per se. Here is an electric field string. Strings closed loop form up and down quarks.

Because the black hole is rapidly spinning and this blueshift occurs in the presence of a super intense magnetic field; the spin direction favors a quantum accumulation of energy forming a positron transformation into a more massive proton. The electron remains leptonic with some hadronic characteristics depending on position and speed.

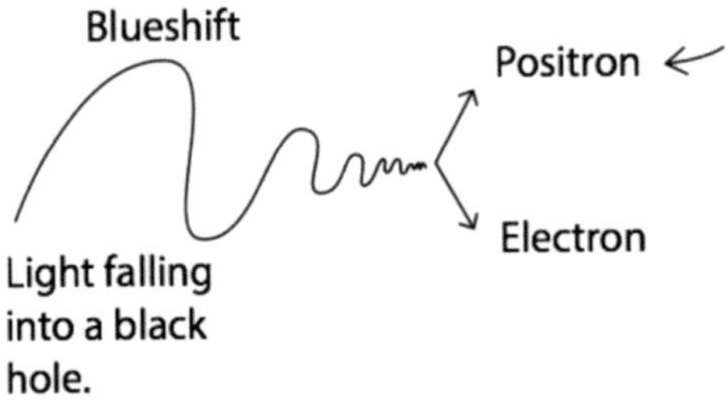

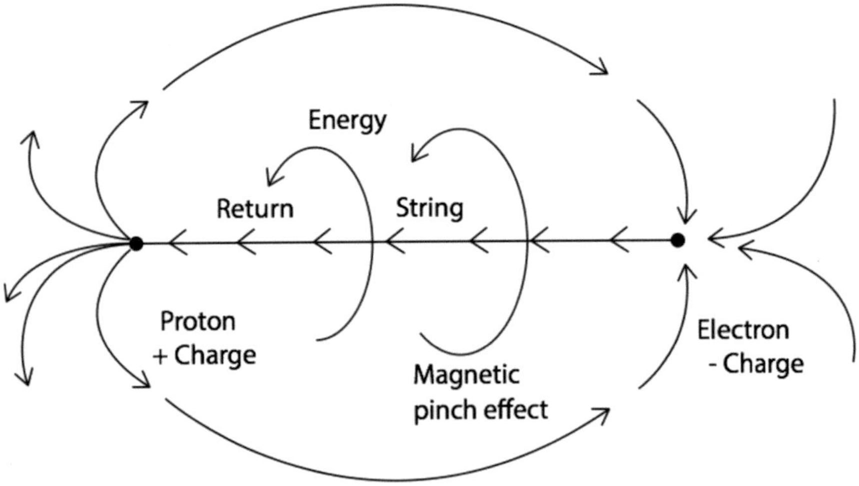

The magnetic field blowout at opposite ends of a black hole is like a repulsive spring for plasma making a black hole a messy eater. At the center of the universe the black hole is so big what falls in is in balance with what falls out - like an automatic valve on James Watt steam engine.

Einstein's Lorentz radical is a fudge number derived from vector trigonometry.
It's goofy

$$e^2 = V^2 + Vx^2$$

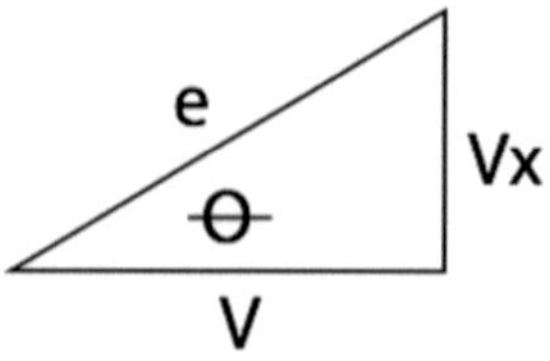

$$\frac{V^2 + Vx^2}{e^2} = \frac{e^2}{e^2} = 1$$

$$\frac{e^2 - V^2}{e^2} = \frac{Vx^2}{e^2}$$

His $M, = \dfrac{Mo}{\sqrt{1 - V^2/e^2}}$ Is what?

$$\sqrt{\frac{e^2 - V^2}{e^2}} = \sqrt{Vx^2/e^2} = Vx/e =$$

$$\sqrt{1 - V^2/e^2} = Vx/e$$

Nuts,
The rest mass doesn't change m, is about the energy conserved it takes to accelerate or decelerate a rest mass where is it conserved?

In the Higgs field which is not mass.

For numbers to not be imaginary - They must refer to something both real and measurable.
Math is a language of numbers. In any language it's possible to tell a fib.

10,000 BC

Poisden Verso Hare Krishna

(Atlantic) (India)

Nuclear War

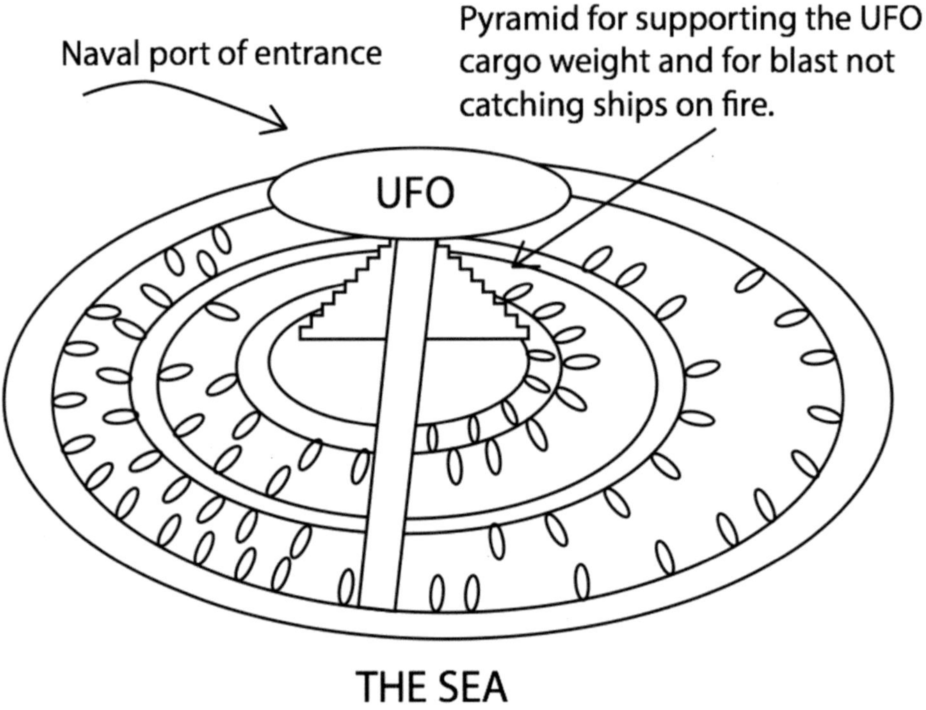

The design of Atlantis makes sense if the horns are removed and a flying saucer rests atop to bring in the cargo. The horns is a corruption of the story. The UFO's took in the merchandise from sails ships in the harbor and transportedto the interstellar spaceship fleet.

One of these off the coast of the equatorial Africa. Another in America. Another in the mediterranean. Several around the world. Plato got his story of Atlantis from the Egyptians which got their story drawn on stone. Notice the similarity of the Plato pyramid and the Aztec. The circular design of the port like that photographed from space. Another under water off the coast of equatorial Africa. That circular port design was to transport cargo to its center. The only way from there is up. Religion corrupted the story and the Aztecs offered sacrifice atop pyramids to lure the alien spaceships back. It didn't work.

HI STEPHEN: I THOUGHT YOU MIGHT LIKE TO ADD THIS TO YOUR BOOK

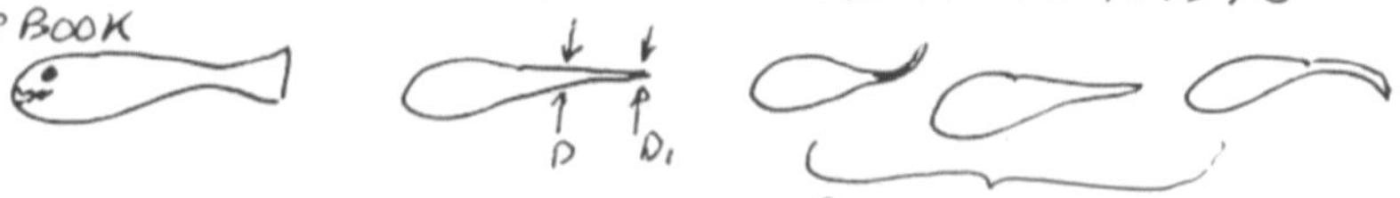

FISH TAIL BENDS WHILE FLAPPING CAUSING FOWARD THRUST. THIS PRINCIPLE IS INCORPORATED IN DRAGON FLY WINGS

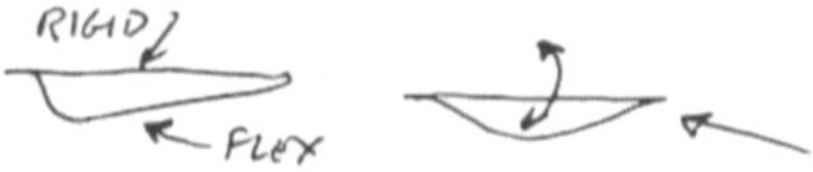

THE CHINESE FAN CAN FOLD FLAT.

WHY NOT ROTATING BLADES? HIGH SPEED FANS ARE DANGEROUS IN CRASHES.

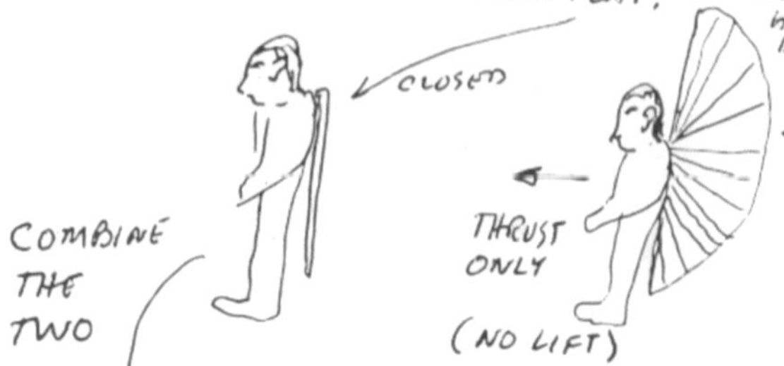

TWO WINGS FLAPPING CHINESE FAN DESIGN MOVES ALOT OF AIR WITH GEAR REDUCTION FOR TORQUE MATCHING. ENGINEERS KNOW HOW TO DESIGN THAT

TWO HALF FANS (CHINESE) LEFT AND RIGHT SHOULDER OPENS LIKE A BIRD WING, NO FLAPPING

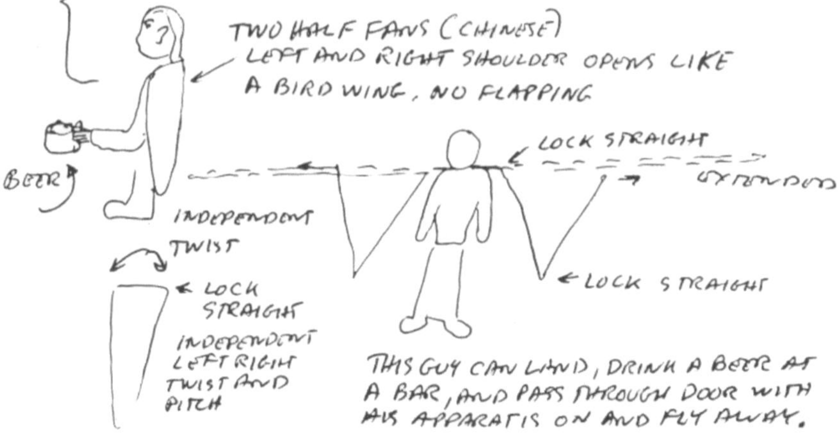

THIS GUY CAN LAND, DRINK A BEER AT A BAR, AND PASS THROUGH DOOR WITH HIS APPARATIS ON AND FLY AWAY.

ADD A LITHIUM RECHARGIBLE BACK PACK BETWEEN THE TWO FLAPPING VERTICAL CHINESE FANS AND THE GUY SHOULD FLY. LEAVE IT TO ENGINEERS TO DETAIL. TO COMBINE THE VERTICAL AND HORIZONAL OPERATION AS IN HOOPING CRANE WINGS IS MORE COMPLICATED AND CAN BE DONE BUT — PROGRESS SHOULDNT TAKE GIANT LEAPS. SOME USE CAN BE OBTAINED OF THIS DESIGN BECAUSE ITS COMPACT ENOUGH TO PASS THROUGH A HOUSE DOOR.

I DAYDREAMED THIS IN HIGH SCHOOL 1960 – 61 BUT NOONE THEN HAD LITHIUM BATTERIES AND LIGHT DC MOTORS AND ELECTRONICS.

MY 1960 DESIGN LESS REFINED

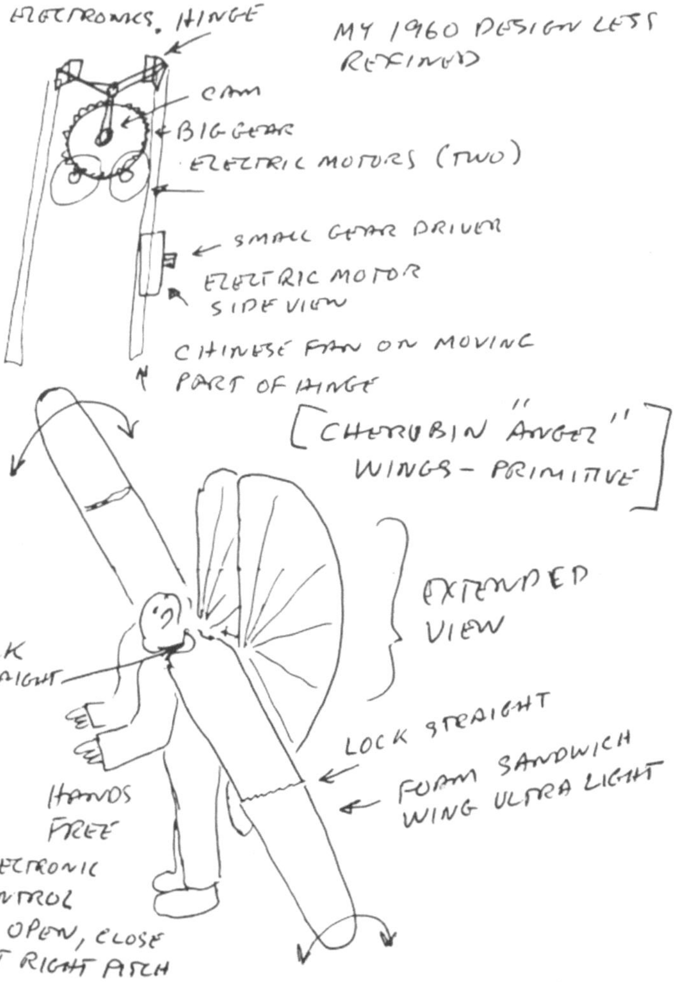

[CHERUBIN "ANGEL" WINGS – PRIMITIVE]

ELECTRONIC CONTROL FOR OPEN, CLOSE LEFT RIGHT PITCH ON/OFF VIBRATION BY SENSORS ON WRIST LEFT & RIGHT HAND AND VARIABLE VIBRATION SPEED

THIS IDEA ORIGIONATED IN THE MIDDLE AGES. NOTICE MY CHERUBIN BOY KISSING HIS FEMALE CHERUBIN MATE. (COUSHIN PICTURE) HE HAS BIRD (HORIZONTAL EXTEND) WINGS, SHE HAS VERTICAL BUTTERFLY (CHINESE FAN) WINGS – IN "KISS" [illegible]

Dene Watson
Inventor

Negative Ion DC Apparatus

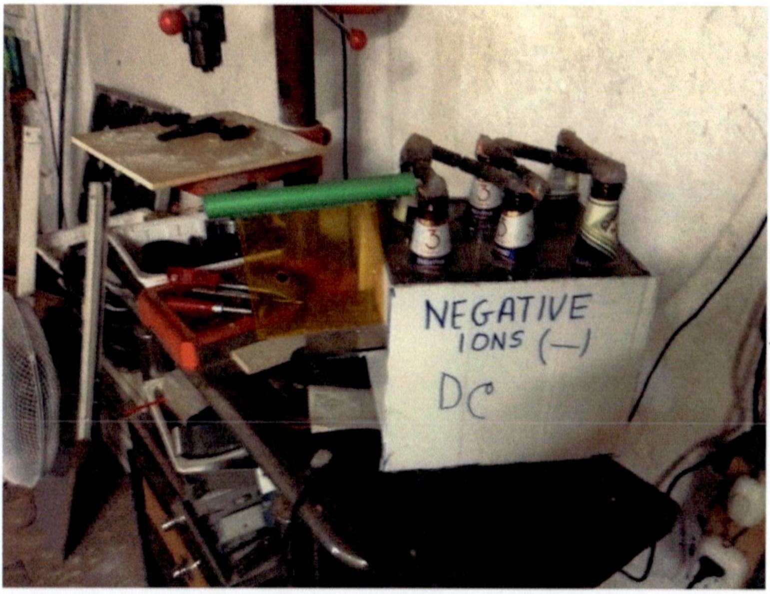

I built something like this back in 1993. It takes in 8,000 volts AC and transforms to a much higher DC output. I'm using the negative ions from that to stimulate hair growth, kill airborne viruses and bacteria, and for DC ionic thrust models.

The beer bottles are high voltage capacitors for the reason home built - electronic stores wont sell stuff reserved for business and research institues.

Every time I have the sniffles, I breath through my nose the negative ions and that stops me from having a flue. It's not easy to build that device because it takes high voltage diodes no longer in stock and a step up transformer hard to find. If it was positive ions I would get a headache and feel sick.

At such low frequency, not nearly as powerful as what I was using in Wylie, Texas.

The Japanese now have a negative ion ionizer built into some electric heaters, very recent on the market.

UFO LAB MOTOR Electromagnetic Wave Plasma Low Frequency

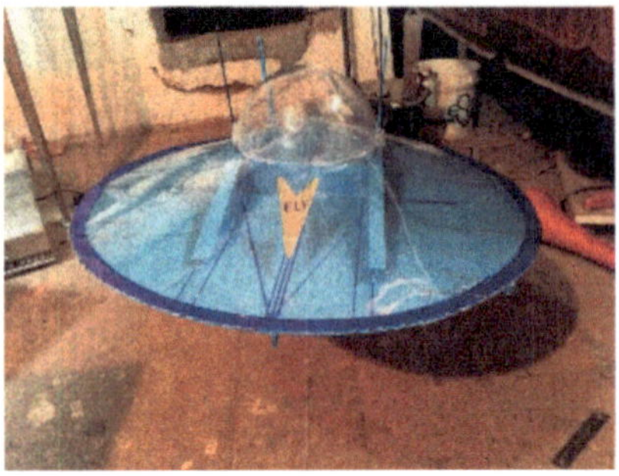

The sustained thrust is straight forward acceleration.
Tested OK. No air intake. No exhaust. No ground effect lift.

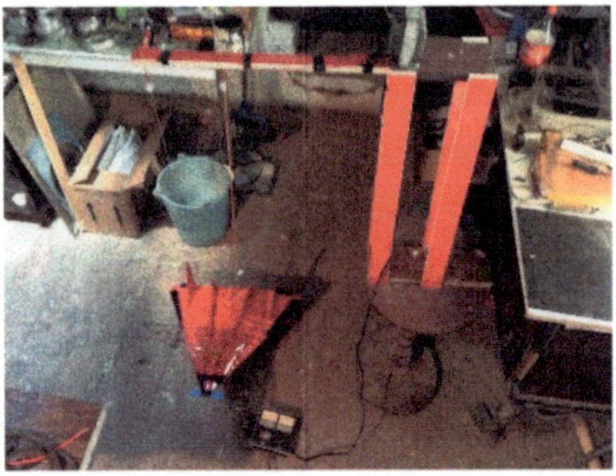

ELF UFO Delta Wing Motor. Works. No air intake No exhaust. Accelerates straight forward.

Tubular Configuration - Faraday Motor
UFO ELF design variant. Works. It's a solid state device. Accelerates straight forward.
The motor is inbedded in the hull itself.

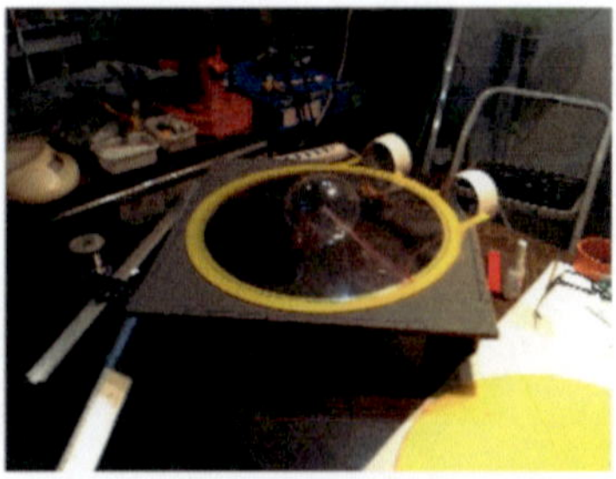

The faraday motor has two configurations; tubular and hull.
I have tested both successfully.

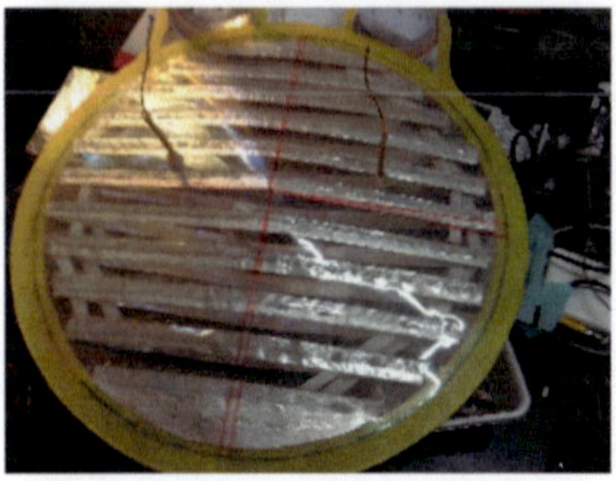

L: Those strips in the picture are functional modified bevatron dees. A motor design predicted by the propulsion matrix I actually built and got a sustained thrust from.

R: Also pictured: AC powered UFO and fluorescent light.

I decided to build a curved hull for external ac ionic thrust comparison.
It's not finished yet.

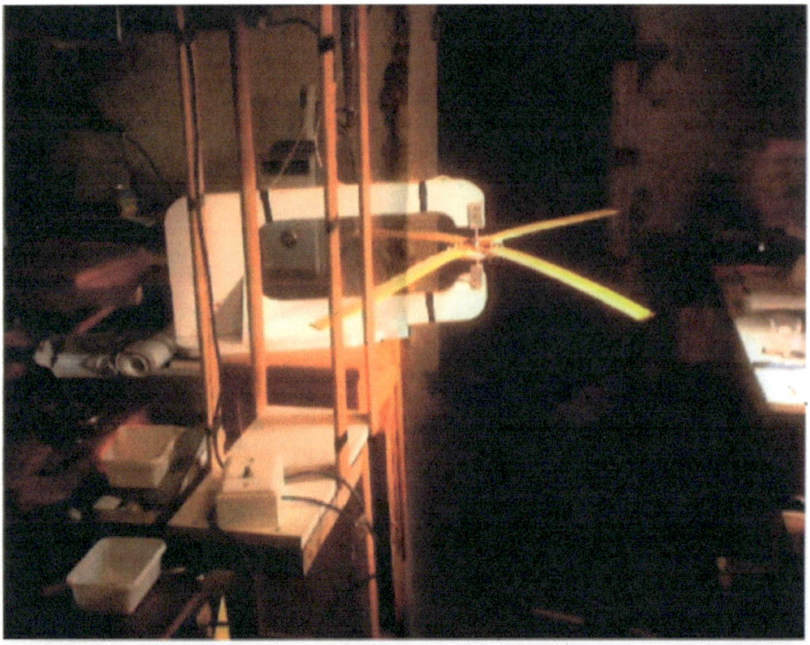

Like my photo? It spins without motors or gear reduction by ionic. This spins fast (ionic helicopter wings).

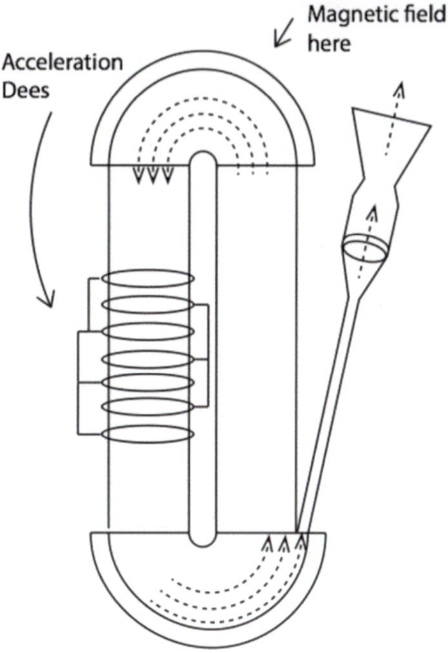

Cyclotron Drive

It's a stack. I drew single for simplicity. It can be built small enough to fit inside a truck.

Remember my DC ionic generator? It generates a wind of negative ions felt by the hand. Reversing the diodes causes the same device to generate positive ions having a wind of equal intensity.

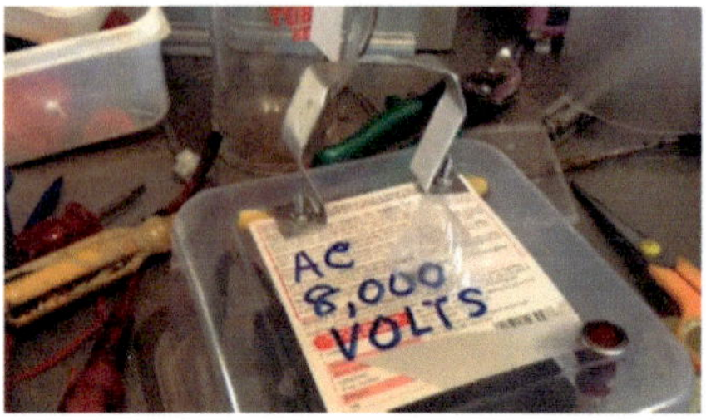

8000 volt AC arc I used as a cigar lighter.

What about AC? This experiment in AC shows an arc with both terminals bare metal.

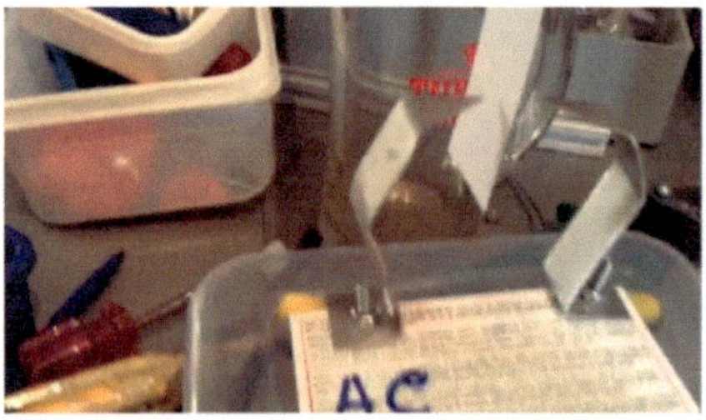

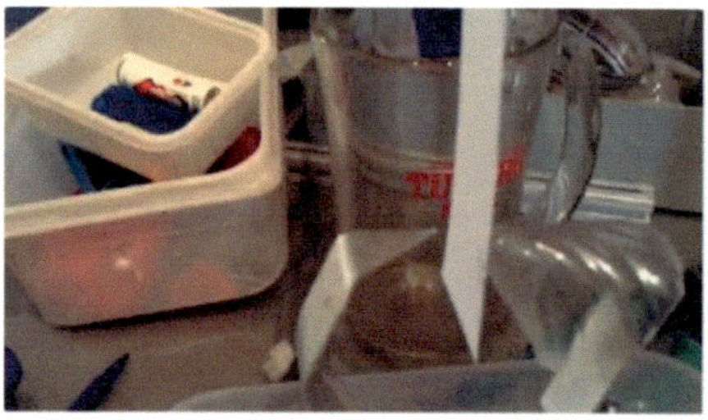

Then I insulate one terminal only.

See the paper being pushed?

Then I put the insulation to the other terminal and the ions push the paper the other way.

Indifferent if the ion waves and changeling from positive to negative many times a second.

This experiment is similar to the one I did two years ago and from that discovery I made my AC flying saucer models work.

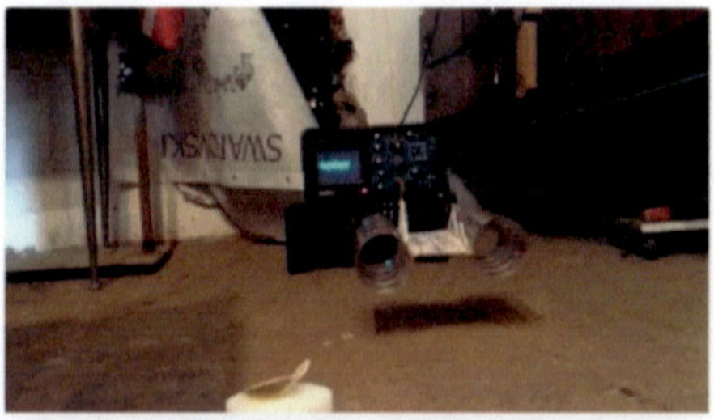

I didn't think to turn the oscilloscope in the last experiment.

As you can see the 50 cycle European AC input the saucer transforms into a higher frequency. There should be harmonics where AC power input converts to linear acceleration even more efficiently.

I don't have lab equipment to raise the applied frequency in search of the best harmonic.

As I said earlier, this conversion is very efficient as my amp meter doesn't move. Increasing the voltage increases the static thrust.

The space separation I am using between nude needlepoint array and insulated aluminum strip is 12 mm (see Page 53). Less than 9 mm the ions burn a short hole through the polyester plastic insulation.

I don't have the instrument to raise the voltage past 10,000 Volts. If the best puncture resistant insulating material is used, the whole saucer would glow and the static thrust much higher by raising AC voltage to just below puncture strength. I know this because I can detect some glow in the dark and the kind of gas influences color.

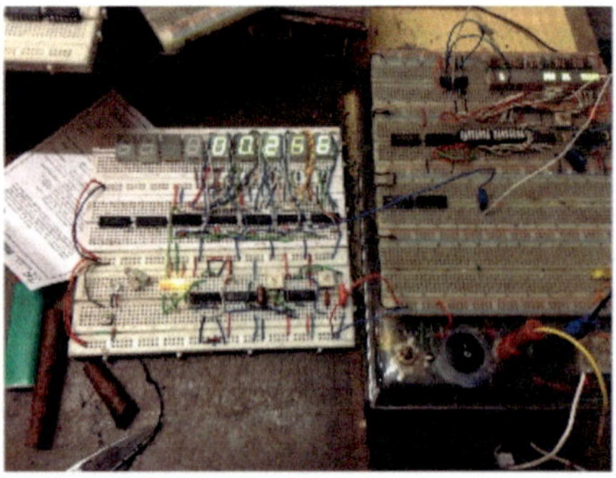

That picture is a frequency counter I designed and built.

It's for practice in connecting a variety of IC chips in my possession to see what they do.

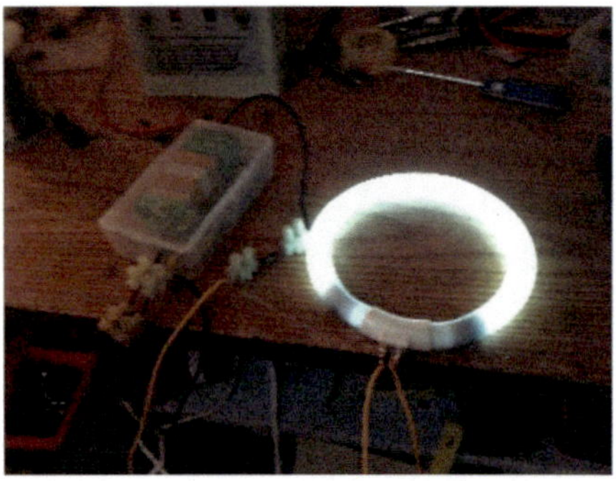

Left: That Made In China microcircuit isn't really what it seems to be. My oscilloscope detected a powerful extremely high frequency but that's not the point.

Right: It's filaments are cold. It was a fluorescent lamp thrown away for that reason. I junked the ballast. The electronic circuit I invented enables the lamp to glow for centuries for as long power is applied. I beat the clowns to their own game.

No filaments that get hot and burn out. Consumers should be treated nice, not cheated. Lamp engineers knew this but pigs out to cheat customers pay them to make cheap consumer products.

See this experiment? I took my long fluorescent tube not connected to anything I turned all the lights off and I held it near the Digicom pocket internet transmitter and the fluorescent lamp glowed briefly by slowly rotating the tube in my hand. Even a meter away from the Digicom. For the glow the angle must be right necessitating slow rotation as if it was a cheerleader baton.

A long fluorescent tube like this one works better because your hand around the tube completes a capacitive circuit with your body and the tube part of the circuit. The microwaves passing vibrate electrons perpendicular to direction of wave travel and in the tube having rarefied gas the kinetic energy released is sufficient for a dim glow by ultraviolet light generated striking the white element coating. With frequency increase electromagnetic waves act less like waves and more like bullets. A kind of lepton to hadron shift that doesn't concern this experiment

An intense microwave signal would cause the lamp light on bright and stay on as was demonstrated at Fort Gordon, Georgia when for laughs some guys pointed a radar at a GI who walked out the storeroom with a bundle of fluorescent lights a mile from the radar. It scared him and he dropped the lamps and ran about a mile away from the radar. The year was 1966.

See this experiment? Lots of intense X-rays are passing through the fluorescent tube as I demonstrated earlier with the Geiger counter on. No flourescent glow. The amount of microwaves intense enough to harm you would make the fluorescent tube glow without it being connected to anything.

If you have this cute looking plasma globe lamp with lightening coming from an inner ball - Get rid of it. It kills people. It's an X-ray generator.

I had one of these balls but I also had a home built Geiger counter. That Geiger Counter saved my life. What harm that ball did was a bad backache as I had it behind my seat.

Needle Array.

Pictured Right: The needle point here is bare metal. The distance between here and here is 12mm. Electric field strength is volts/distance

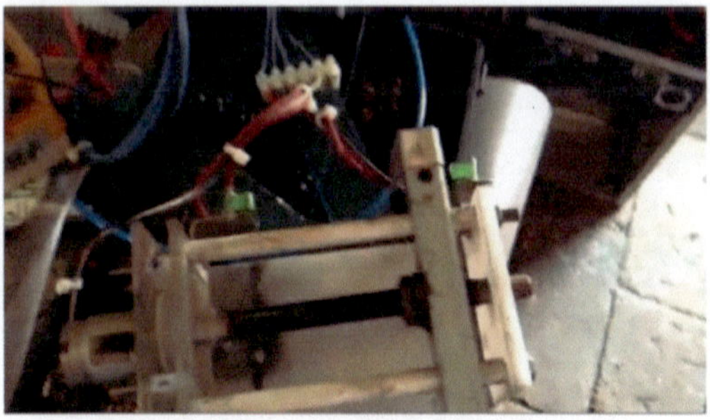

The electronic equivalent to James Watt's automatic valve.

Brushless DC Motor.

DARPA wanted someone to invent a brushless DC motor.
As you can see, I went beyond that and eliminated the magnetic windings too.

Electric Propeller Driven Wind Bike

Lab Notes

Question: On the Faraday motor tubular configuration demonstration. What causes the buzzing sound?

Answer: The buzzing or frying egg sound is charged molecules breaking the sound barrier.

Question: Ion lifters are criticized as being too weak.

Answer: Nonsense. The atmospheric ionic motors I demonstrated are weak because I got the transformers from the junkyard designed for other applications. Those kinds of motors help engineers design the toroid bevatron. That thruster can generate lethal high accelerations that would squash an astronaut like a bug under foot which is why I emphasized fail safe design to prevent too many protons from being released at once. Ohm's Law says in a series DC circuit the ampere reading is the same. The ampere is derivative how many charged particles pass a reference point per sec. That's like comparing people walking in parallel and in single file. The result? In a copper wire a lot of electrons in parallel passing per second must equal the number of electrons single file in the arc passing per second for the ampere to be the same. The speed the electrons reach across one cm arc at 8000 volts potential is enough to generate X-rays near single file because of the magnetic pinch effect while in a wire their speed in parallel is about one meter per hour. That's why get such a high X-ray reading. My Geiger counter reads high.

1970s Little Rock, Arkansas (Page 1)

The ten foot saucer of 1971 I built a special tow for it and I started building it under the supervision of Dr. Bronco Arkansas Polytechnic College and finished it in Little Rock, Arkansas. Look on the map the distance between those two places. He [Dr Bronco] had a doctorate degree in physics. I built some of the parts that went into the saucer with the university lathe and he taught me their Wang Computer for calculations. His propulsion motor theories were ahead of the times and we discussed some of them (See Pages 4 and 6).

Wylie, Texas (See Page 2)

This is proof I was in correspondence with DARPA when Wylie ordered the destruction of my four yard wide flying saucer. On the list DARPA then sent me they wanted someone to invent was a brushless DC motor. As you can see, I went beyond that and eliminated the magnetic windings too (see Page 53). For the reason DARPA was searching to lay their hands on then AC motor I now have, and then US Embassy in Rome ignored my announcement two years ago I had one, it's probable DARPA found another talented individual before or after Wylie's interference in my attempt to discover that motor. When the City of Wylie had me detained in jail to have my four yard wide saucer cut to pieces and taken to Flash Gordon's junkyard. What the City of Wylie didn't know before that incident a representative of one of the richest person in Dallas, Texas who later ran for president of the United States personally inspected my flying saucer. I was registered in Collin County as a business that went by the name "VESCIO RESEARCH." I was in correspondence with the military agency called "DARPA" and they wanted me to discover how to build the AC ionic motor I now have, but then didn't. I was very close to making that discovery. To do that it was first necessary to build a DC ionic thruster seen in this picture (Page 32 and Page 43). I had to use DC to discover how positive ions compare with negative ions before moving on to an AC device and I was doing that in the Wylie tornado shelter days before I was arrested. The four yard wide saucer served the purpose of cleaning ground obstacles with ducted fan liftoff to see if I could push the saucer forward on ionic motor thrust for DARPA to see. What DARPA was really trying to do was to discover the most talented individuals in the country in this area of coming up with innovative ideas that served the country. The City of WYLIE acted like the mob who sought to hang Leonardo da Vinci for his parachute invention.

The Sachse, Texas Saucer (Page 3)

This is a better flying saucer design for several reasons. The flat screen for looking down is superior to a glass window for night landing because electronics makes special cameras see better. Electronics gives automatic stability not necessitating the rotors be above center of weight. The location of the forward fans permits the saucer to angle like an airplane wing cutting down on fuel consumption. The large intake cuts down on fuel consumption. The forward fans can have hand grids so nobody on the ground can stick their hands in. The forward fans are hydraulic motors giving smoother speed control. The lift fans are planetary gear counter rotating eliminating undesired gyroscopic effects.

Propulsion Matrix (Page 17)

I drew all the motors in the propulsion matrix in the sense of representative example. The United States after confiscating Nazi jet airplanes spent in the 20 year successive more than the development of the atomic bomb to improve the Nazi jet engine. That's just one motor in the propulsion matrix. The rocket jet motor predicted by my propulsion matrix is under construction by the British [Skylon]. It goes orbital without stages and returns not as a glider but with powered landing like a jet. The motor cools intake air for higher compression up to 26 miles altitude with the liquid hydrogen radiator then past 26 miles draws on its internal lox to orbit. All this makes into an interesting book that can be copyrighted and sold. You can computer draw the rough sketches. If the United States wants to keep that a secret or make a false claim I disclosed their secret research, they will have a tough time explaining how I acquired all those motor designs without so much as setting foot in Area 51. I gave the ex mayor of Grand Prairie, Texas the propulsion matrix a little after you was born [1975]. From that it is not difficult to come up with the boron glass bevatron. It may have been included. Then also I gave copies to an American Indian where I worked at Sunrise Industries and these people are practiced in business administration. It doesn't surprise me if some of those motors got secretly researched by the American Skunk Works with my name and identity dropped. This partly explains some UFO Sightings as an attempt by the Defense Department to perfect some of those motors. I remember going to the Navy base near Mountain Creek Lake and handing over to a high ranking Navy official the cluster rocket ramjet in the 1970's motor design (see Page 5).

Rejected & Stolen

No matter what I invent, I make no money because in every case the idea gets rejected in my face and stolen behind my back. Bell Helicopter in Hurst, Texas told me 40 years ago the only outside offer was from the founder of Bell Helicopter and threw my flying car design in the trash can about as advanced as the Sachse, Texas drawing (Page 3). We were both snubbed. Dr. Bronco was smarter than me. Then the Russians their superior pump snubbed by NASA. Dr. Ruby (Carlo Rubbia) of Italy also snubbed by NASA for his americium nuclear rocket. Big companies want to steal from others or originate their own designs without paying anyone except their own employees. The smartest intellectual in the United States that topped Einstein was assassinated in the 1930's for being too smart. His name? Tesla.

I had an appointment with Vought Aerospace in 1977 and after I showed them a simple ionic thruster, they walked out without listening to the rest of what I had to say. Here in Italy after demonstrating an ionic motor sponsored by Dalbe Sky Cranes (former employer) in his reception room, I handed over a copy of the propulsion matrix to some German Engineers around 1998 and never heard from them again.

Giving Ideas Away

In answer to your question, why did I give the designs away. Look at the big picture. Man is born to be wild-not in the criminal behavior sense. We are civilized Klingons where savage blood is not Conan the Barbarian but to rise to the challenge of the survival of the species in a savage universe that tries to kill us off. Those designs for having been given away survived Wylie, Texas madness to strip me of everything I owned. I knew this solar system is dirty with asteroids and comets and it only takes one big event to set man back to the chimp. So by sacrificing my own monetary interest I contributed to advancing research into advanced propulsion motors in the interest of the survival of the human race in the hope the hardware would be developed in time to have man self-supporting on more than one celestial body.

All my advanced propulsion motor as generalized designs are free. I don't want even a penny from anyone. Organizations that transform generalized sketch of that nature are really the inventors of their product and if the government wants to keep something secret, it's in the detail that belongs to the producer, not in my generalized sketches. The neat thing about generalized designs is the impossibility of back tracking who had that idea first. Look at the art in Latin America that can be seen only from the air. Who saw it from the air? A big kite can lift a man by a rope tied to a stake. Then after the wind gives altitude he let's go of the rope and glides down. There are thousands of inventions that were not invented first in Europe or America but lost to history and re-invented.

Here in Italy it costs 50 thousand euros to patent something simple. That only gives the Chinese an advantage. They buy production machinery and flood the markets with finished products with no regard for patent laws outside China. Einstein worked in the United States Patent office as a German citizen during Hitler's rise to power. It was legal spying but because Hitler treated him like shit, he became a real American.

NASA Rejection

Years ago Russia declared to NASA a new pump design. NASA rejected it. Russia responded by transporting their rocket with its new pump to Cape Kennedy in Florida and NASA launched it. It went orbital on less fuel consumption.

Large Chemical Burn Rockets

If I sent you a variety of advanced propulsion motor designs, my emphasis on large chemical burn rocket boosters doesn't seem to make sense. Actually it makes a lot of sense by reason the chemical rocket motor is well developed. For this reason the chemical burn rocket motor is in the hands of the major countries and there is an undeclared but very real race off this planet. The United States does not have the luxury of time. The large chemical burn booster rockets are needed to set up industrial facilities on the moon and Mars while the other motors take the back burner for their time will come. Otherwise the iron hand will be set up that stops any rocket launch from succeeding if one country obtains too big a lead over the others in space (illustrated in SDI-Star Wars no access to space without permission of the dominant fleet). An American astronaut suggested some 30 years ago an orbit to orbital reusable shuttle that would pick up and drop off the landers. That shuttle would slowly rotate for artificial gravity. The idea got rejected. For the reason NASA rejected that astronauts idea to have an orbit to orbit spaceship that rotates for artificial gravity the moons of Mars are needed to provide shade against the solar

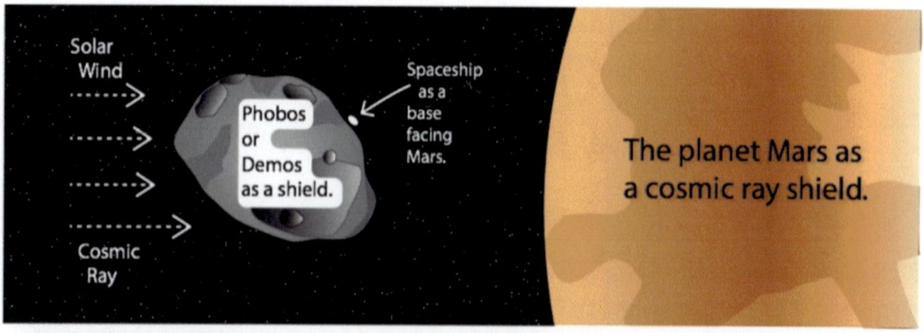

wind. The problem is not for the astronauts that land but the ones still in orbit and remain in orbit for months. The solar wind charged particles come to an abrupt stop by collisions with the spaceship hull. That results in X-rays that penetrate inside the spaceship if it is small. I demonstrated that by putting a Geiger Counter near a 8000 volt AC arc I used as a cigar lighter. If it is a big rotating spaceship in Mars orbit, no significant radiation problem and Mars moon shadowing not needed. What many don't realize is what makes land on Mars valuable and the moon is people's feet. Can you walk on it and construct cities, etc? More briefly accessibility. If so government money backed by industrialized real estate etc. causes the space program to operate at an profit eventually. A women that gets pregnant in a weightless environment and remains weightless all the way to delivery-the baby won't live. To stay in space a long time requires taking the astronaut seriously

The undeclared race for Mars is not like Neil Armstrong 1969 first on the moon and after a few more missions the congress canceled the program. Russia isn't backing down as happened in the race for the moon. ESA is also developing that capability and China is trying to catch up from behind. Why? Because access to real estate has monetary value translated into backing monetary currency with titles to real estate, mining operations, industry, etc. like Alaska has more value now then when Russia sold the land to the Americans. It's a land grab bigger than the gold rush to California. All that is dependent on means of transport or space travel hard ware. The real value is not in my rough sketch motors but translation into detail. Rough sketch is trail blazing.

Nuclear Fusion For Space Travel

The nuclear fusion reactor in Southern France [ITER] is the size of a cube the height, width, and length of a four story building. The concept is very simple and a crude sketch a child in elementary school could draw, but the devil is in the detail. The breakthrough was the earlier break even reactor built and tested a decade ago using ordinary copper wire coils. The results permitted engineers to predict the substitution of superconducting coils resolves the problem but doing that isn't easy. The engineers think they have the answer and that is what this reactor is out to demonstrate but won't reveal the details. Greedy capitalist don't want a cheap substitute for petroleum because of investments in the petroleum industry. With a nuclear fusion reactor generating 5 megawatts, there is an embarrassing choice of chemical fuel that can be made. All this is interrelated. Big boosters, energy source past Mars, high ISP rockets, an asteroid defense shield, and the eventual evolution of interplanetary spaceships into interstellar. All this greedy capitalist are against. They want their profits today by bribing politicians and by impeding new technologies that would compete with petroleum. Update: The nuclear fusion that uses laser power in California (Lawrence Livermore National Laboratory) was successfully tested on 14 February 2014. It generates more power than is used to fire up the lasers. In 1970 Dr. Bronco proposed the idea in a discussion with his physics students and afterwards somebody built it. The toroid nuclear fusion reactor is behind because of funding cuts and is more complicated. Either can be used for interstellar travel. The biggest problem is keeping the interstellar spaceship warm. In interstellar space without heat - air turns liquid. It takes several centuries to make either nuclear fusion dependable enough to attempt interstellar spaceships.

Fission Vs Fusion

The radioactive waste for nuclear fusion reactors is virtually nonexistent and well managed. In contrast nuclear fission for an interstellar spaceship would be a nightmare. Two deuterium fusion collisions create one stable helium atom not radioactive. Fission is chain reaction, messy and dangerous. Fusion is clean, not chain reaction (in the sense controlled fusion reactor for electrical power where chain reaction is excluded).

Deuterium & Tritium

Deuterium has one neutron. Tritium has two neutrons. A free neutron transforms back to a proton by releasing its electron in about 15 minutes. Radioactivity has to do with nuclear instability by more or less neutrons that what the stable element has in its nucleus. At the heavy end of the periodic table, the bigger the nucleus, the closer in the electrons shells leading to instability. An earth sized moon of Saturn has methane oceans. It's not feasible to transport that methane here. With deuterium and tritium that's different. The amount of hydrogen gas that would fill a 4 foot weather balloon made of deuterium or tritium hydrogen in nuclear fusion reactor generates about 20 megatons hydrogen bomb except that the reactor is designed to nuclear fuse that amount of hydrogen gradually generating 5 megawatts continually night and day. The correct word is not helium 3 but tritium. Deuterium is a hydrogen atom with a neutron. Tritium is a hydrogen atom with two neutrons in the nucleus. A neutron is a proton with an electron smeared on its surface in mesonic bond exchange. To make the long story short most hydrogen atoms have no neutrons but those that do extracted from water is fuel for nuclear fusion reactors like the one being built in Southern France. It was supposed to go into operation in 2012 but funding cuts is delaying until 2018. That is a big thing in space. It takes a nuclear fusion reactor for interstellar travel because of the abundance of hydrogen in ice low gravity worlds, comets, Oort clouds. On the moon the trapped ice is richer in deuterium, tritium than water here on earth, may also be the case for Mars. With superconducting coils replacing conventional coils, the toroid nuclear fusion reactor is expected to generate from 10 to 50 times the energy that is used to make it work.

UFOs And Science

Question: Does antigravity exist described in UFO sightings?

Answer: Any alien UFO accelerating faster than 8G is drone. Gravity is not space curvature but an energy attractive field. The existence of energy in a void doesn't affect the void in any way. They can't star gate.

UFOs and Protons

A proton traveling at 90 percent light speed has a kinetic energy storage similar to a antimatter particle at rest only that it's much safer to manipulate protons. Does that explain why some UFO phenomena has to do with electricity? Charged particles in motion is current flow from the perspective of ohms law. Ohms law doesn't care if it's electrons or protons in flow.

Remember the oval UFO without intake or exhaust holes whose entire surface is smooth and no windows? It uses for propellant mass less than one gallon of drinking water like what you get from the water faucet. With that it can accelerate vertically at 8g with a manned crew and arrive at the Oort cloud in only a few days. In 1942 above Los Angeles one of these allowed allegedly if the report is correct, to be bombarded by anti aircraft fire without returning fire as a test of its ability to stop any hostile fire bullet or shell. Why? Because in interstellar space that capability is used to blaze a clear trail ahead of the interstellar fleet. I know how that motor works and I'm not talking.

SETI: SETI did pick up some alien transmissions but did not know how to identify it because scientists still push the dumb idea that the aliens are only a few thousand years more advanced and live on some goldilocks planets. The nearest being 12 light years from earth. The authentic signal when they pointed the telescope was not from any stellar system but from interstellar space and SETI dismissed it. The aliens were not intentionally trying to contact us but transmitting to another of their own spaceship computer talk and the line of transmissions for a brief period was pointed in our direction. The aliens use an electron bevatron with electrons passing through a magnetic culmination for highly directional transmissions not interceptive as the communication is in between interstellar spaceships in space and we don't know how to design a receiver even if we could put the receiver between two interstellar spaceships. Lower frequencies is rarely used. It makes sense interstellar spaceships exchange information by directional transmissions to check on the integrity of data banks to assure repairs on such things as their nuclear fusion power source is defect free carried out mostly by robotics. The last thing the aliens would want to attempt is an omnidirectional transmissions saying we are here. SETI should spend more time dreaming what they would expect to find on an interstellar spaceship. Their concept of finding an earth like planet in the beginnings of space travel is 1930 outdated Flash Gordon fiction. There are other planets in this galaxy with civilizations in the beginnings of the development of space travel hardware but none close enough for SETI radio interception. It hould have been obvious to Drake the more advanced civilizations have explored farther out from the planet of their ancestral origins and hence more probable to have discovered this planet first, but for being more advanced, more offended by the criminality on this planet hence more elusive toward us. The aliens don't want to send us messages. Their contacts are informal. They don't want to establish formal contacts with so many criminal minds on this planet. They detest the idea of enslaving the masses and are not out to take advantage of anyone. They have no intentions conquering this planet and no intentions of interfering with man's technological progress. SETI is founded on the pseudoscientific beilief the aliens wish to establish contacts with us with old outdated communications like Marconi radio. Scoffers at interstellar tavel won't consider the obvious. The aliens concetrated on perpetual maintenance after perfecting nuclear power sources and advanced propuslion devices for their interstellar spaceships. Their interstellas spaceships is their permanent home. They are not seeking to take possesion of another earth like planet in another stellar system before they can refuel or make repairs. The alien spaceship averages 30 kilometer diameter and inside is a lot activity like what goes in New York City to keep everything operating. What it is for is not to go from sollar system to another stellar system in less than 40 years, but for its inhabitants to exist and live normal lives. They not only have to keep their spaceship in repair but also have to pass on the technology to successive generations like in a big city. It's all centered around the nuclear fusion reactor. They have several inside the spaceship. In interstellar space the temperature outside the walls drops a few degrees above absolute zero. The ITER experiment in southern France is important. Man has to harness nuclear fusion energy or go the way of the dinosaur.

(YOU DONT HAVE TO INCLUDE THIS IN THE BOOK – ITS SO YOU CAN UNDERSTAND THE PROBLEM)

WHAT A MONKEY BRAIN IS: EGO MANIA, POWER ATTRACTS EGO MANIACS

A MONKEY THAT LOVES HIMSELF TO THE EXCLUSION OF ANYBODY ELSE IN FRONT OF A MIRROR SEES HIS OWN REFLECTION AS ANOTHER MONKEY (SOME REAL MONKEYS ARE THAT STUPID) APPLY THE NEO-NATZI MENTALITY TO THAT. THE MONKEY TRIES TO BULLY AND DISCOVERS HIS OWN REFLECTION TRIES TO BULLY – SO THEN THAT SPARKS A FIGHT WITH THE MONKEY THROWING PUNCHES. THE END RESULT IS TRAJEDY.

Greedy capitalist don’t understand because a selfish mind is a monkey in his banana tree that looks out only for himself (Page 66 illustration) and that is why monkey’s don’t build cities.

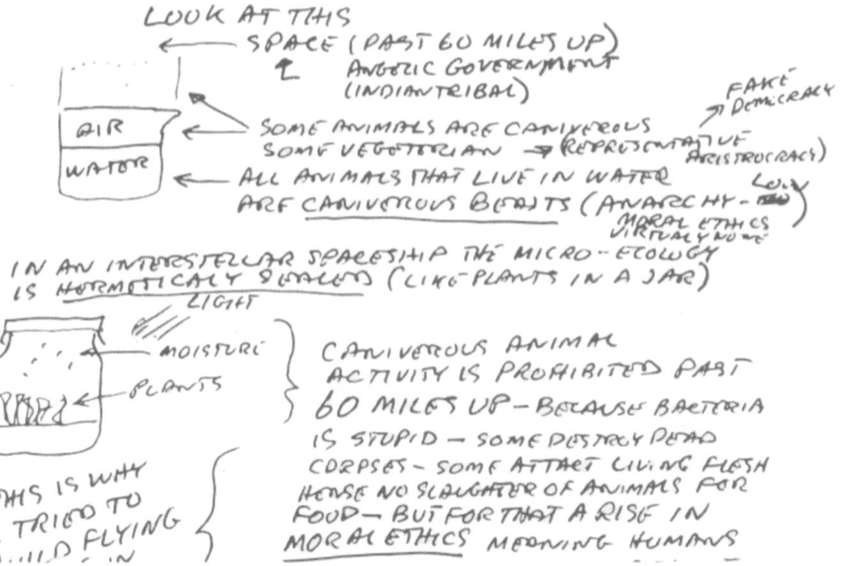

The aliens are not ethically perfect. You have to imagine yourself living on a interstellar spaceship with 500 other families. This is an indian tribe without a monetary system with a chief that distributes work and no one does without. It’s higher ethics and not what the European invaders demonstrated when they stepped of the Mayflower in Jamestown. The aliens inforce a prime directive not on us but themself to protect themselves. We are brain size behind them and their advanced technology in the hands of some earth ego maniac means interstellar spaceships that shoot at peaceful aliens like when Columbus planted the Spanish flag on the western hemisphere and claimed all Native Americans savages to rob and exploit. That's coming from a really low mentality. Aliens don't want to shake hands with us, much less share their advanced technology. The only way aliens will trust humans is for us to catch up with them enough and that may take many centuries. Besides the alien interstellar spaceship in the solar system’s

Oort cloud intends to leave after completing repairs to be replaced by another. They don't find humans all that interesting.

Ancient Aliens (Page 40)

In this scenario the Aliens took some humans with them as far back as 8,000 BC. The humans multiplied in interstellar spaceships with the capacity to launch an invasion to reclaim this planet. The original intent was to save the species from extinction by growing another human culture off this planet. I have no information wether the human culture off this planet plan to launch an invasion. I don't exclude the possibility but it could be very minimal.

US Space Command

I'm one of the founders of the United State Space Command in the sense the idea was quite popular among intellectuals and I suggested the idea to the government. (I don't claim to be the first to have that idea it was during the Ronald Reagan admin).

The Secrecy Paradox

Nobody has come up to me representing any government to tell me to keep secret any of my motor sketches. If they had, I would have kept secret. It's a kind of paradox. Keeping a device secret does not guarantee with 7 billion people on this planet someone else doesn't come up with the same idea. Apparently the United States government doesn't know how to deal with this problem other that making arrests on bogus charges as happened to me and that didn't work. If they had asked me not to reveal this or that I would have complied.

Angels

Biological beings normally after they die including animals get recreated biologically without past life memory. There exists a large but limited number of angels given perpetual existence and among them cherubim. The cherubim are the children of seraphim and a higher standard of moral ethics is taught to them by their spiritual parents than what ordinary humans practice. Some humans become cherubim. Their number is limited. Put another way, some not all human animas become cherubim animas.

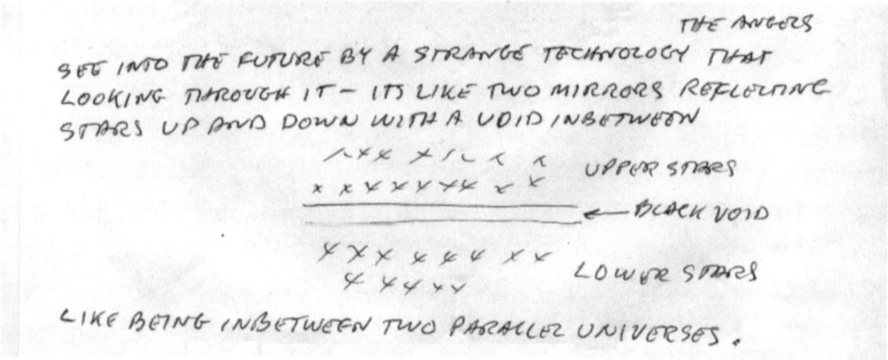

All this talk about aliens and dream fantasies (see Page 26 and Page 35) is because the mind asleep spends some eight hours a day in sleep and nature causes people to dream for a purpose. To me it's like a Donald Duck comic book where the emphasis is not content but coincidence. By accident here and there wrapped in fiction something true may exist and that has helped me invent things, but mostly stay out of trouble.

Angel Wings

My experiments in flapping wings (see Page 42-43) suggest it is possible to design a foldable flapping wing that would enable people to fly. It would be powered by lithium rechargeable batteries and suitable DC motors gear reduction matched. Something like the pelican that takes off with a running start and land with a running speed - It takes expensive production equipment and computer design enhancement to optimize performance.

More on Wylie, Texas (Page 2 and Page 56)

At the same time I had a big flying saucer in Wylie another private citizen owned his own Chinook military helicopter and tested it with cables tied to stakes on his own private residence. I saw it hovering 30 feet above his yard as he tried to learn its controls. The Chinook was in very bad shape for having been bought from a military scrapyard or auction. Wylie didn't bother him. Then there was a small club that pooled their donations and built a hovercraft. Wylie didn't bother them either. Wylie picked on me because some nut set three houses on fire, one next to ours, and mistook me for that man. I was detained in jail 57 days and when the real criminal was arrested the McKinney police announced it by intercom to all the cell members where I was detained and released me, but wanted me to plead guilty on a bogus charge so I wouldn't take them to court on false arrest. Orville had a one grade education and Wilber a third grade when the Kitty Hawk flew certainly not from Wylie. I drank beer with a space shuttle astronaut in Pensacola, Florida at the same time Wylie was looking to arrest me for 10 cents.

More on Wylie, Texas (Pages 32, 43, 56)

It would be in violation of national security to make public that particular circuit. I keep it like a key that fits a lock. Without it engineers don't know how to design a lightweight powerful AC generator that that gets driven by a jet or rocket burn canister staged axil flow turbine. They could build a powerful AC generator alright without knowing the correct power match oddities that requires their AC generator to be designed in a way that differs from conventional AC generators. And that circuit says why. I believe the US Skunk Works can discover that secret on their own after several failed attempts to correctly power that kind of motor. That's why I'm not talking and have only weak models to demonstrate.

Nuclear "Cold" Fusion Ball

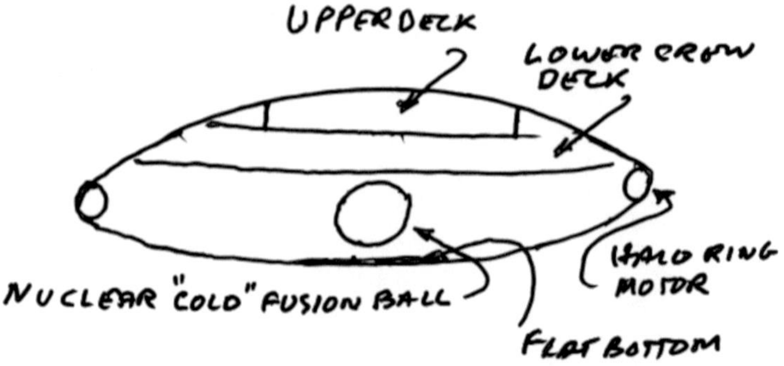

What the Russians have they showed on Russia Today was a segment equivalent (primitive) of what corresponds to the alien "cold" nuclear fusion ball - not as theory, but as functional hardware. The aliens or "seraphim" showed me in a dream their "fusion ball" as a cross sectional segment. Did the Russian device (not a ball) work? Yes, but it was built for taking measurements for engineers to have data they need to produce one that puts out more energy than it consumes. Actually seraphim can not manipulate matter but can read the minds of both humans and aliens so then what I saw in the dream was not a detailed abstract of an alien device they saw somewhere in the galaxy. I understood the Russian scientist knew how to design a primitive device independently of correspondence with me and before I had the same idea - making my abstract drawing worthless - so I destroyed my design (worthless in that Russian scientist know much better than me even before I had the idea how such a device would work (It's not a bomb)).

1977 Engine (See Page 19)

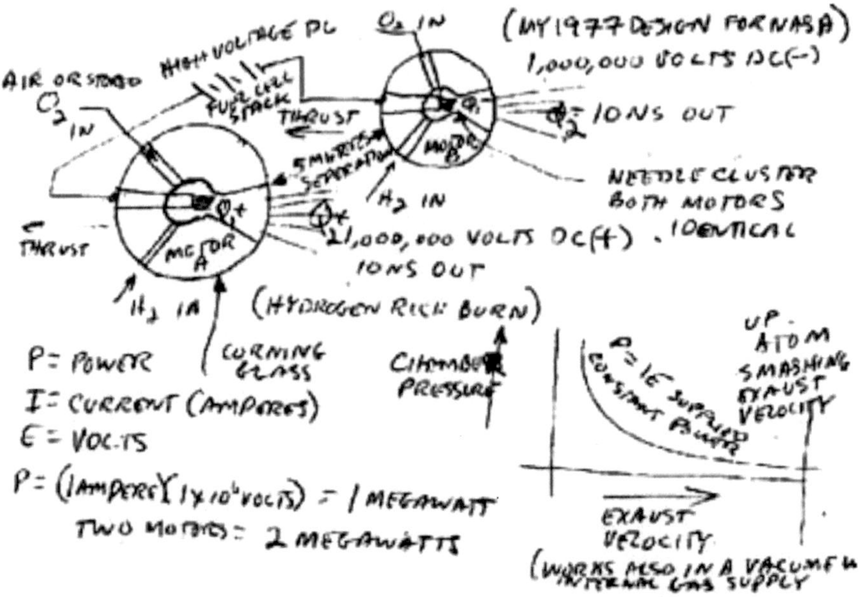

Original Low Res Image.

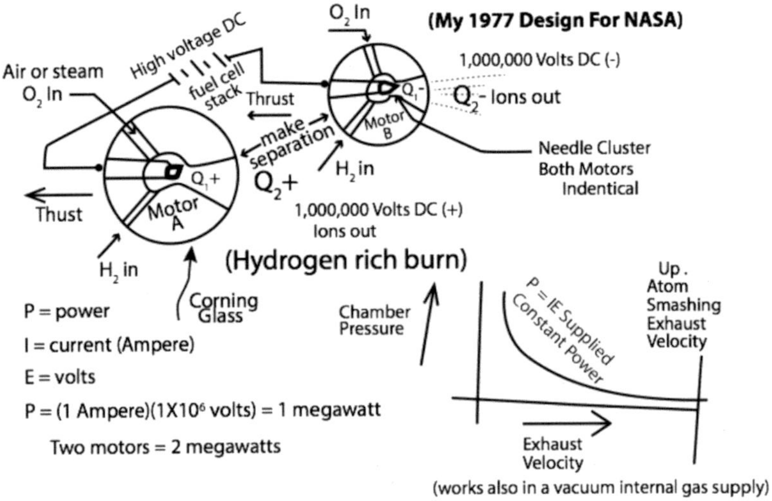

Redrawn Version.

One Segment Nuclear Cold Fusion Deuterium Reactor

(For Nuclear Fusion Toroid Reactor see Page 29)

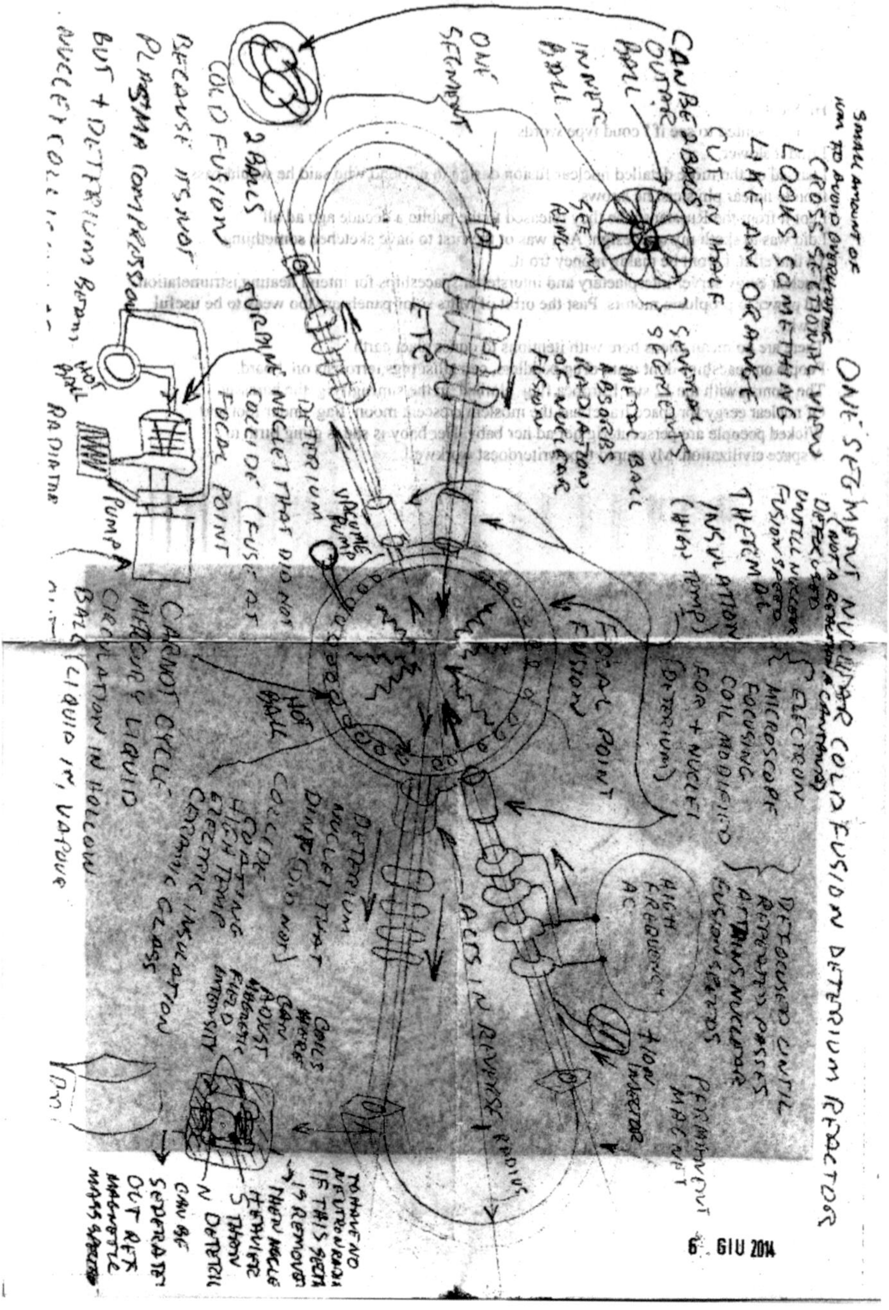

No Plasma Cold Fusion

THIS IS NOT RUSSIAS DESIGN BUT MY OWN PRIOR TO MY SEEING RUSSIA'S RELEASED TO THE PUBLIC IN THE LATE 1990'S OR EARLY 2000'S. MY OWN ORIGIONAL I DESTROYED BUT CONATE POZZOLO HAS A COPY OF THE ABSTRACT (GENERAL DESIGN) I DESTROYED. I DONT REMEMBER THE EXACT DATE (COULD HAVE BEEN AS EARLY AS 1995)

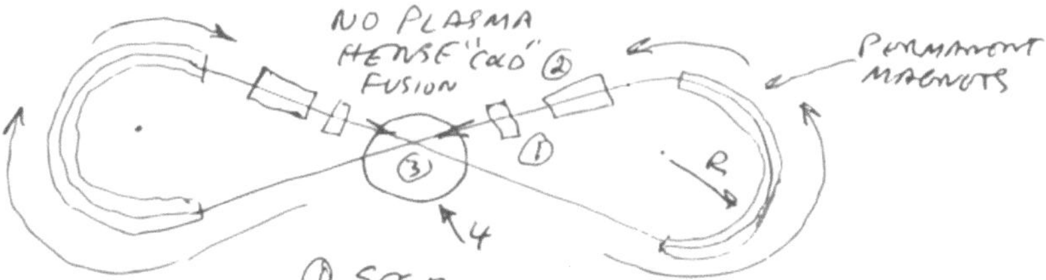

(CROSS SECTIONAL VIEW) ONE SEGMENT

1 SEE ELECTRON MICROSCOPE TECHNOLOGY

THE SAME IN THE X Y Z CO-ORDINATES RESULTS IN A SPHERE (OUTER) COLD. INNER SPHERE HOT

2 SEE CHERN LINEAR PARTICLE ACCELERATOR FUNCTION AT REDUCED SPEEDS

3 FOCAL POINT OF DETERIUM PARTICLE COLLISION

4 RADIATION ABSORBING INNER BALL (BOILS LIQUID MERCURY WHOSE VAPOUR DRIVES ELECTRIC GENERATOR TURBINE) (STAGED AXIL FLOW)

ITS NOT AN ATOM SMASHER. ITS FEW MILLION ELECTRON VOLTS POTENTIAL FUSES DETERIUM NUCLEI FORMING HELIUM WITH RADIATION ENERGY RELEASE — HENSE THE PERMANENT MAGNETS SERVE ONLY TO CHANGE THE DIRECTION FOR REPEATED PASSES OF THE DETERIUM PROTON BEAM, FOR THIS REASON THE CURVE RADIUS (R) IS MUCH SMALLER THAN FOR AN ATOM SMASHER. RUSSIA'S IS SUPERIOR IN THAT THEY DEMONSTRATED THEIR SEGMENT FULLY FUNCTIONAL — NOT AN ABSTRACT DRAWING ON A PIECE OF PAPER — MEANING THEY HAD THIS IDEA MUCH EARLIER THAN WHEN I SKETCHED MINE IN THE MID OR LATE 1990'S. COULD THIS HAVE BEEN SKETCHED BY NUCLEAR PHYSICISTS AS EARLY AS THE 1950'S YES, ITS TOO SIMPLE AMOUNTING TO A SPEED REDUCTION OF AN ATOM SMASHER TO EFFECT FUSION INSTEAD OF NUCLEAR FRAMMENTATION

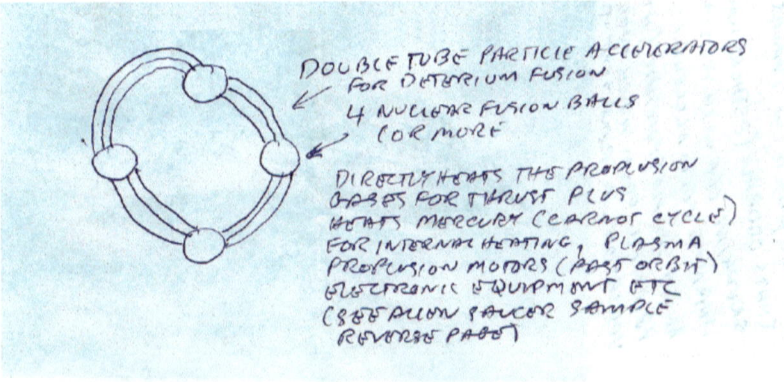

Double tube particle accelerators for deuterium fusion.

Four nuclear balls or more.

Directly heats the propulsion gasses for thrust plus heats mercury (carnot cycle) for internal heating, plasma propuslion motors (past orbit) electronic equipment etc. (see alien saucer sample reverse page)

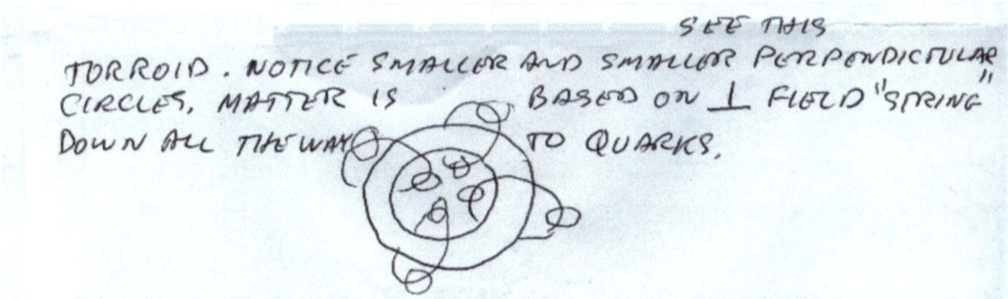

See this toroid? Notice smaller and smaller perpendicular circles. Matter is based on ⊥ field "spring" down all the way to quarks.

"Forbidden Planet" Flying Saucer Analogy (see Page 33)

For lunar/martian landing; initial lift-off aerospike/ plug nose can substitute or supplement as seen in the fantasy flying saucer model "Forbbiden Planet."

Vertical ascent/decent illustrated here tubular AC ionic motors to cushion landing in touchdown phase or initial vertical ascent. AC applied voltage in excess of a hundred thousand volts. Internal gas supply rocket turbine exhaust supplying torque power to the AC high voltage generator. The horizontal ionic thrusters off in the saucer hull surface. Makes sense?

AC/DC
adjective
alternating current/direct current.

aerodynamic
adjective
of or relating to aerodynamics: aerodynamic forces.
• of or having a shape that reduces the drag from air moving past: the plane has a more aerodynamic shape.

aerospike engine
noun
The aerospike engine is a type of rocket engine that maintains its aerodynamic efficiency across a wide range of altitudes. It is a member of the class of altitude compensating nozzle engines. A vehicle with an aerospike engine uses 25–30% less fuel at low altitudes, where most missions have the greatest need for thrust.

afterburner
noun
an auxiliary burner fitted to the exhaust system of a turbojet engine to increase thrust.

americium
noun
the chemical element of atomic number 95, a radioactive metal of the actinide series.(Symbol: Am)
Americium does not occur naturally and was first made by bombarding plutonium with neutrons. It has been used in industrial measuring equipment as a source of gamma rays.

americium-242m
The isotope Am242m is an interesting nuclide as it shows one of the known highest neutron fission cross section. This could be useful in some special applications, like the nuclear reactor for space propulsion proposed by C. Rubbia.

ampere (abbr.: **A**)
noun
a unit of electric current equal to a flow of one coulomb per second.

anima |ˈanəmə|
nounPsychology
Jung's term for the feminine part of a man's personality. Often contrasted with animus (sense 3). the part of the psyche that is directed inward, and is in touch with the subconscious. Often contrasted with persona.

antigravity
nounPhysics
a hypothetical force opposing gravity.

Atlantis
a legendary island, beautiful and prosperous, which sank into the sea.

bevatron
noun
a synchrotron used to accelerate protons to energies in the billion electron-volt range.

Carnot cycle
noun
is a theoretical thermodynamic cycle proposed by Nicolas Léonard Sadi Carnot in 1824 and expanded by in the 1830s and 1840s. It can be shown that it is the most efficient cycle for converting a given amount of thermal energy into work, or conversely, creating a temperature difference (e.g. refrigeration) by doing a given amount of work.

celestial
adjective [attrib.]
positioned in or relating to the sky, or outer space as observed in astronomy: a celestial body.

centrifugal force
nounPhysics
an apparent force that acts outward on a body moving around a center, arising from the body's inertia.

CERN (European Organization for Nuclear Research)
noun
is a European research organization whose purpose is to operate the world's largest particle physics laboratory.

chassis
noun (pl. same)
the base frame of a motor vehicle or other wheeled conveyance.
• the outer structural framework of a piece of audio, radio, or computer equipment.

Cherub |ˈCHerəb|
noun (pl. cherubim) |ˈCHer(y)əbim|
a winged angelic being described in biblical tradition cas attending on God. It is represented in ancient Middle Eastern art as a lion or bull with eagles' wings and a human face, and regarded in traditional Christian angelology as an angel of the second highest order of the ninefold celestial hierarchy.

Corning Glass
Corning incorporated is an American manufacturer of glass, ceramics, and related materials, primarily for industrial and scientific applications.

coulomb (abbr.: **C**)
nounPhysics
the SI unit of electric charge, equal to the quantity of electricity conveyed in one second by a current of one ampere.

cyclotron
nounPhysics
an apparatus in which charged atomic and subatomic particles are accelerated by an alternating electric field while following an outward spiral or circular path in a magnetic field.

da Vinci, Leonardo
(1452-1519)
was an Italian inventor, artist, architect, and scientist.

Dee
noun
semicircular electrodes, called dees from their shape: each is a D-shaped half of a pillbox.

deuterium
nounChemistry
a stable isotope of hydrogen with a mass approximately twice that of the usual isotope.(Symbol: D) Deuterium atoms have a neutron as well as a proton in the nucleus, and the isotope is present to about 1 part in 6,000 in naturally occurring hydrogen. It is used as a fuel in thermonuclear bombs, and heavy water (D 2O) is used as a moderator in nuclear reactors.

diode
noun Electronics
a semiconductor device with two terminals, typically allowing the flow of current in one direction only.
• a thermionic tube having two electrodes (an anode and a cathode).

Drake equation
Astronomy
a speculative equation that gives an estimate of the likelihood of discovering intelligent extraterrestrial life in the galaxy. It was formulated by the US astronomer Frank Drake in 1961.

electron gun
noun Physics
a device for producing a narrow stream of electrons from a heated cathode.

electron volt(abbr.: **eV, MeV**)
noun Physics
a unit of energy equal to the work done on an electron in accelerating it through a potential difference of one volt.

emu
abbreviation Brit.
• electric multiple unit.
• electromagnetic unit(s).

entropy
noun
1 Physics
a thermodynamic quantity representing the unavailability of a system's thermal energy for conversion into mechanical work, often interpreted as the degree of disorder or randomness in the system.(Symbol: S)
2 lack of order or predictability; gradual decline into disorder: a marketplace where entropy reigns supreme.

equilibrium
noun (pl. equilibria)
a state in which opposing forces or influences are balanced: the maintenance of social equilibrium.

fusion
noun
• Physics short for nuclear fusion.
Fusion power is a primary area of research in plasma physics.

G
abbreviation
a unit of gravitational force equal to that exerted by the earth's gravitational field.

Gay-Lussac's law
Chemistry
a law stating that the volumes of gases undergoing a reaction at constant pressure and temperature are in a simple ratio to each other and to that of the product.

Geiger counter (also Geiger-Müller counter)
noun
a device for measuring radioactivity by detecting and counting ionizing particles.

ground
noun
electrical connection of a circuit or conductor to the earth.

ground effect vehicle (GEV)
noun
is one that attains level flight near the surface of the Earth, making use of the aerodynamic interaction between the wings and the surface known as ground effect.

H II
noun
ionized atomic hydrogen, pronounced H-two by astronomers (an H I region being neutral atomic hydrogen, and H2 being molecular hydrogen)

H2
noun
Liquid hydrogen

hadron
nounPhysics
a subatomic particle of a type including the baryons and mesons that can take part in the strong interaction.

helium 3
noun
the chemical element of atomic number 2, an inert gas that is the lightest member of the noble gas series.(Symbol: He)

Hubble's law
Astronomy
a law stating that the redshifts in the spectra of distant galaxies (and hence their speeds of recession) are proportional to their distance.

hull
noun
the main body of a ship or other vessel, including the bottom, sides, and deck but not the masts, superstructure, rigging, engines, and other fittings.

hypersonic
adjective
relating to speeds of more than five times the speed of sound (Mach 5).

Infrared
adjective
(of electromagnetic radiation) having a wavelength just greater than that of the red end of the visible light spectrum but less than that of microwaves. Infrared radiation has a wavelength from about 800 nm to 1 mm, and is emitted particularly by heated objects.

interstellar
adjective
occurring or situated between stars: interstellar travel.

International Thermonuclear Experimental Reactor (ITER)
is an international nuclear fusion research and engineering project, which is currently building the world's largest experimental tokamak nuclear fusion reactor adjacent to the Cadarache facility in the south of France.

/sp, ISP, Scientific Impulse
is a way to describe the efficiency of rocket and jet engines. It represents the force with respect to the amount of propellant used per unit time.

joule, joul|(abbr.: **J**)
noun
the SI unit of work or energy, equal to the work done by a force of one newton when its point of application moves one meter in the direction of action of the force, equivalent to one 3600th of a watt-hour.

Large Hadron Collider
noun
The Large Hadron Collider (LHC) is the highest-energy particle collider ever made. It was built by the European Organization for Nuclear Research (CERN) from 1998 to 2008.

Lawrence Livermore National Laboratory
noun
a US federally-funded research laboratory where nuclear fusion experiments are being carried out using lasers.

lepton
noun Physics
a subatomic particle, such as an electron, muon, or neutrino, that does not take part in the strong interaction.

lithium
noun
the chemical element of atomic number 3, a soft silver-white metal. It is the lightest of the alkali metals.(Symbol: Li) Lithium has important links to nuclear physics.

Lorentz force
nounPhysics
the force that is exerted by a magnetic field on a moving electric charge.

lox (also LOX)
noun
liquid oxygen.

Mach |mäk|(also **Mach number**)
noun
the ratio of the speed of a body to the speed of sound in the surrounding medium. It is often used with a numeral (as Mach 1, Mach 2, etc.) to indicate the speed of sound, twice the speed of sound, etc.

Marconi, Guglielmo |mär kōnē|
(1874–1937), Italian electrical engineer; the father of radio. In 1912, Marconi produced a continuously oscillating wave, essential for the transmission of sound. He went on to develop shortwave transmission over long distances. Nobel Prize for Physics (1909), shared with Carl Braun.

Mountain Creek Lake
is a reservoir located 8 miles (13 km) southwest of downtown Dallas, Texas (USA) in the Mountain Creek area of the city. The defunct Naval Air Station Dallas (Hensley Field), now an Armed Forces Reserve Center, is located on the northern shore of the lake. Mountain Creek Lake Park is on the lake's southeastern and south-western edges.

pinch effect
nounPhysics
the constriction of a plasma through which a large electric current is flowing, caused by the attractive force of the current's own magnetic field

planetary gear(also **planetary wheel**)
noun
see sun-and-planet gear.

plug nozzle
noun
The plug nozzle is a type of nozzle which includes a center body or plug around which the working fluid flows. Plug nozzles have applications in aircraft, rockets, and numerous other fluid flows.

positron
nounPhysics
a subatomic particle with the same mass as an electron and a numerically equal but positive charge.

power plant
noun
an installation where electrical power is generated for distribution.
• an engine or other apparatus that provides power for a machine, building, etc.

proton
nounPhysics
a stable subatomic particle occurring in all atomic nuclei, with a positive electric charge equal in magnitude to that of an electron, but of opposite sign.

radioelement
noun
a radioactive element or isotope.

ram air
noun technical
air that is forced to enter a moving aperture, such as the air intake of an aircraft.

ramjet
noun
a type of jet engine in which the air drawn in for combustion is compressed solely by the forward motion of the aircraft.

Rubbia, Carlo
Prof. Carlo Rubbia of CERN (Nobel Laureate in Physics, 1984) brought up the novel concept of utilizing the highly energetic fragments produced by nuclear fission to heat a gas; the extremely high temperatures produced would enable faster interplanetary travel.

rarefied (also **rarified**)
adjective
1 (of air, esp. that at high altitudes) of lower pressure than usual; thin.

redshift (also **red shift**)
nounAstronomy
the displacement of spectral lines toward longer wavelengths (the red end of the spectrum) in radiation from distant galaxies and celestial objects. This is interpreted as a Doppler shift that is proportional to the velocity of recession and thus to distance. Compare with blueshift.

rest mass
nounPhysics
the mass of a body when at rest.

semiconductor
noun
a solid substance that has a conductivity between that of an insulator and that of most metals, either due to the addition of an impurity or because of temperature effects. Devices made of semiconductors, notably silicon, are essential components of most electronic circuits.

seraph |ˈserəf|
noun (pl. seraphim |ˈserəˌfim| or seraphs)
an angelic being, regarded in traditional Christian angelology as belonging to the highest order of the ninefold celestial hierarchy, associated with light, ardor, and purity.

Skylon (spacecraft)
noun
is a design for a single-stage-to-orbit spaceplane by the British company Reaction Engines Limited (REL), using SABRE, a combined-cycle, air-breathing rocket propulsion system, potentially reusable for 200 flights.

steady state
noun
an unvarying condition in a physical process, esp. as in the theory that the universe is eternal and maintained by constant creation of matter.
The steady state theory postulates that the universe maintains a constant average density, with more matter continuously created to fill the void left by galaxies that are receding from one another. The theory has now largely been abandoned in favor of the Big Bang theory and an evolving universe.

string
noun
a hypothetical one-dimensional subatomic particle having the dynamical properties of a flexible loop. • (also cosmic string)(in cosmology) a hypothetical threadlike concentration of energy within the structure of space-time.

subsonic
adjective
relating to or flying at a speed or speeds less than that of sound.

sun-and-planet gear
noun
a system of gearwheels consisting of a central wheel (a sun gear or sun wheel) around which one or more outer wheels (planet gears or planet wheels) travel.

supersonic
adjective
involving or denoting a speed greater than that of sound.

tokamak |ˈtōkəˌmak, ˈtäk-|
nounPhysics
a toroidal apparatus for producing controlled fusion reactions in hot plasma.

toroid
nounGeometry
a figure with a shape resembling a torus.
• Electronics a coil shaped like a torus or doughnut.

torque
nounMechanics
a twisting force that tends to cause rotation.

traction impulse
noun
traction, or tractive force, is the force used to generate motion between a body and a tangential surface, most generally through the use of dry friction, though the use of shear force of the surface is also commonly used.

tritium
nounChemistry
a radioactive isotope of hydrogen with a mass approximately three times that of the common protium isotope.(Symbol: T)
Discovered in 1934, tritium has two neutrons as well as a proton in the nucleus. It occurs in minute traces in nature and can be made artificially from lithium or deuterium in nuclear reactors; it is used as a fuel in thermonuclear bombs.

tungsten
noun
the chemical element of atomic number 74, a hard steel-gray metal of the transition series. It has a very high melting point (3410°C) and is used to make electric light filaments.(Symbol: W)

ultraviolet
adjectivePhysics
(of electromagnetic radiation) having a wavelength shorter than that of the violet end of the visible spectrum but longer than that of X-rays.
• (of equipment or techniques) using or concerned with this radiation: an ultraviolet telescope.

Watt, James
(1736–1819), Scottish engineer. Among his many innovations, he greatly improved the efficiency of the Newcomen steam engine, which was then adopted for a variety of purposes. He also introduced the term horsepower.

Wright Brothers
The Wright brothers, Orville (August 19, 1871 – January 30, 1948) and Wilbur (April 16, 1867 – May 30, 1912), were two American brothers, inventors, and aviation pioneers who are credited with inventing and building the world's first successful airplane and making the first controlled, powered and sustained heavier-than-air human flight, on December 17, 1903.

X-ray (also **x-ray** or **X ray**)
noun
1 electromagnetic radiation of high energy and very short wavelength (between ultraviolet light and gamma rays) that is able to pass through many materials opaque to light.

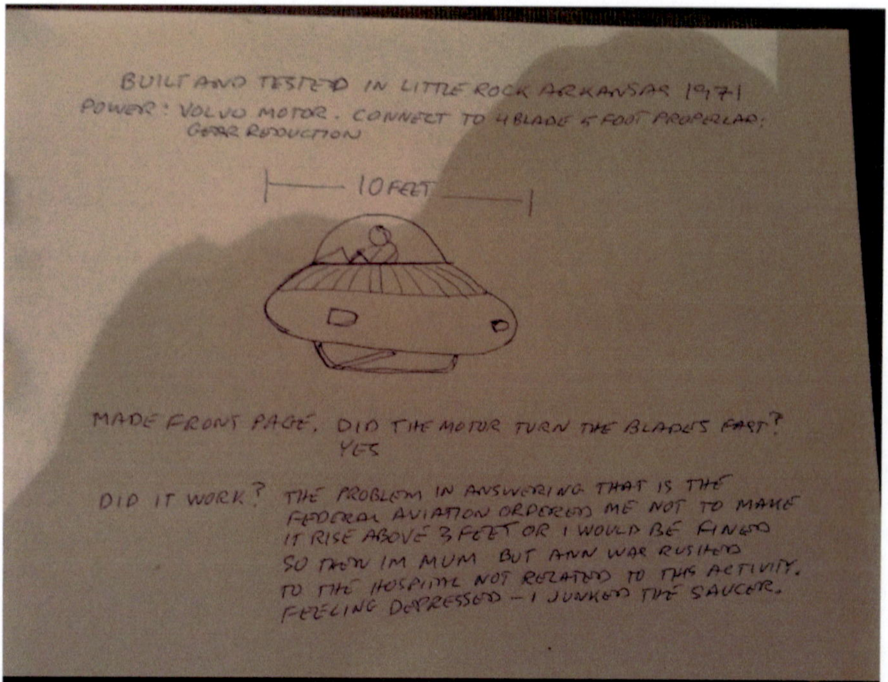
BUILT AND TESTED IN LITTLE ROCK ARKANSAS 1971
POWER: VOLVO MOTOR. CONNECT TO 4 BLADE 5 FOOT PROPELLAR; GEAR REDUCTION
10 FEET
MADE FRONT PAGE. DID THE MOTOR TURN THE BLADES FAST? YES
DID IT WORK? THE PROBLEM IN ANSWERING THAT IS THE FEDERAL AVIATION ORDERED ME NOT TO MAKE IT RISE ABOVE 3 FEET OR I WOULD BE FINED SO THEN IM MUM BUT ANN WAS RUSHED TO THE HOSPITAL NOT RELATED TO THIS ACTIVITY. FEELING DEPRESSED - I JUNKED THE SAUCER.

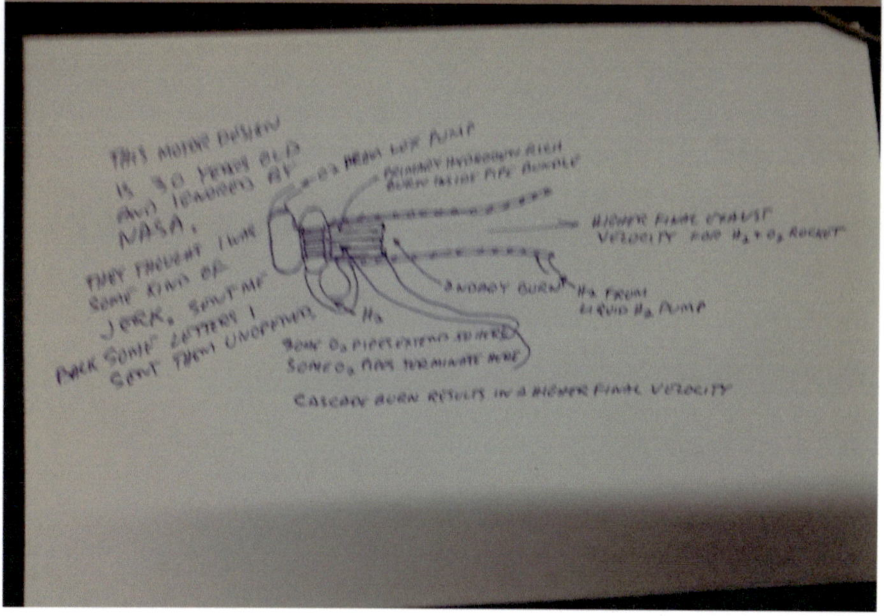
THIS MOTOR DESIGN IS 30 YEARS OLD AND IGNORED BY NASA.
THEY THOUGHT I WAS SOME KIND OF JERK, SENT ME BACK SOME LETTERS I SENT THEM UNOPENED
O_2 FROM LOX PUMP
PRIMARY HYDROGEN RICH BURN INSIDE PIPE BUNDLE
HIGHER FINAL EXAUST VELOCITY FOR $H_2 + O_2$ ROCKET
SECONDARY BURN
H_2 FROM LIQUID H_2 PUMP
H_2
SOME O_2 PIPES EXTEND TO HERE
SOME O_2 PIPES TERMINATE HERE
CASCADY BURN RESULTS IN A HIGHER FINAL VELOCITY

Document Scan - Wylie & Sachse, Texas Sketch

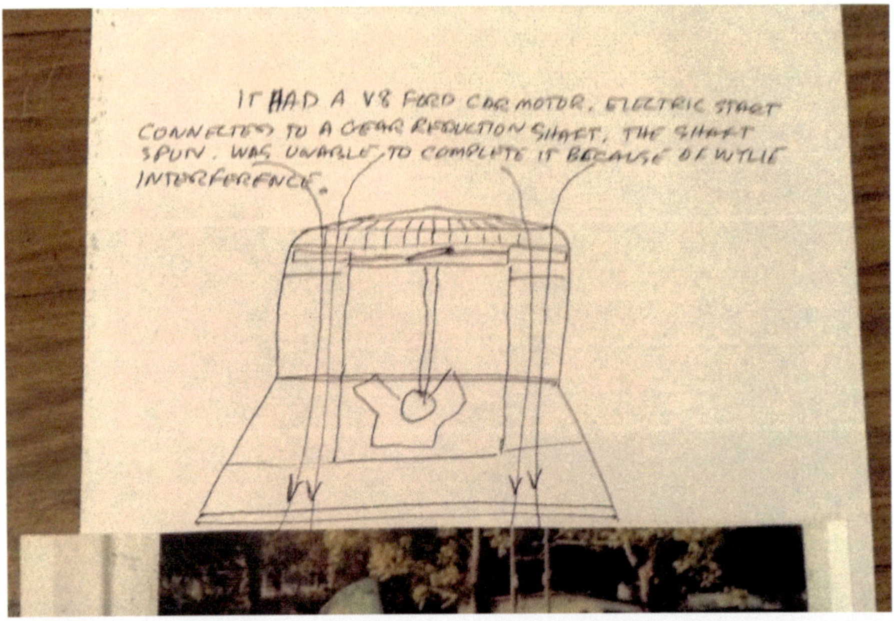

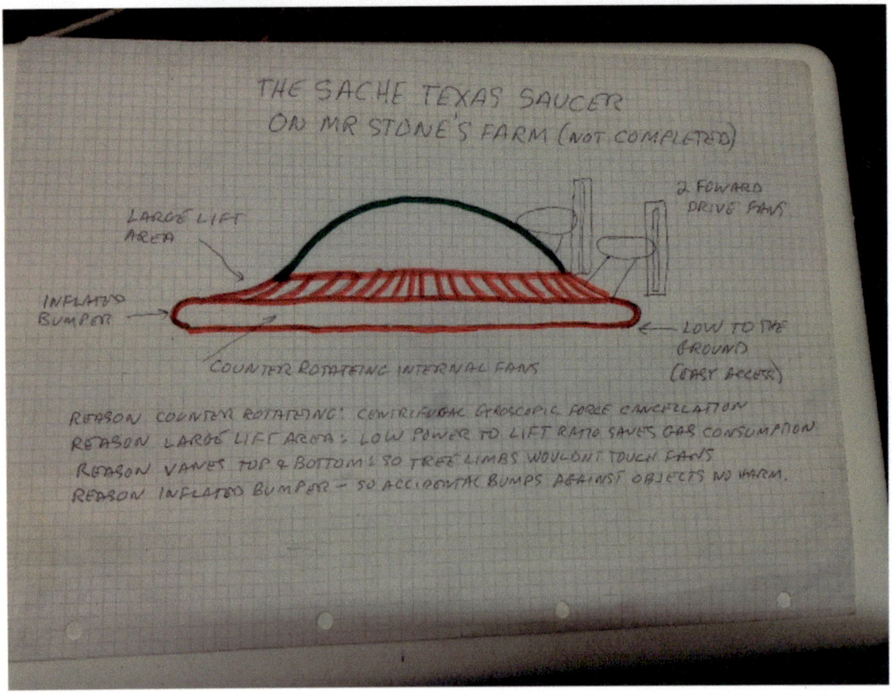

Document Scan - Rocket Array & Geometric Configuration

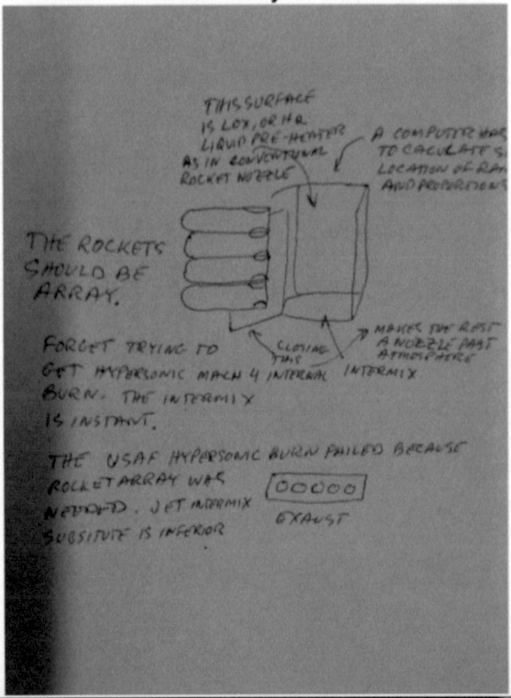

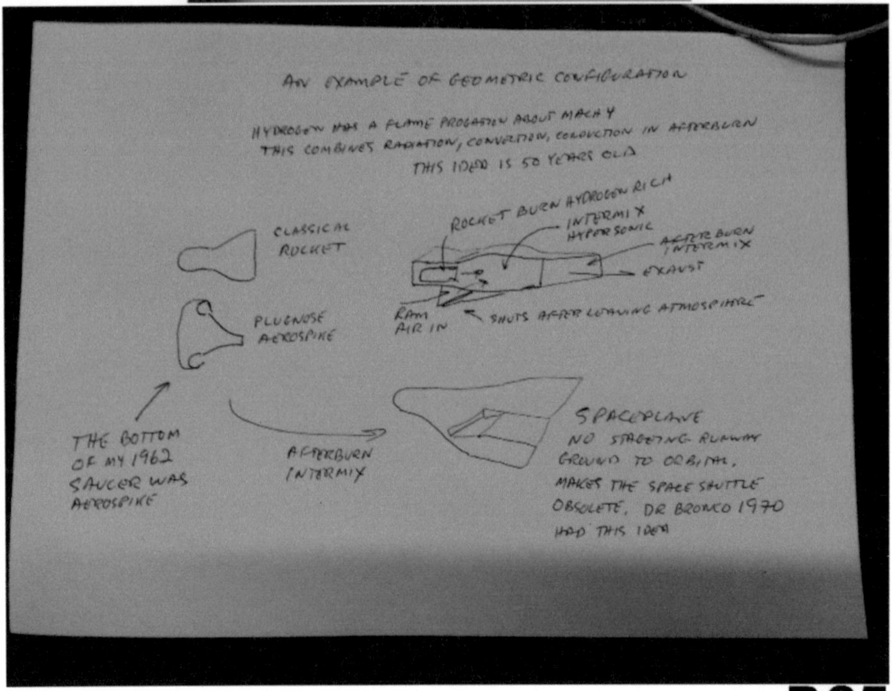

Document Scan - Shell Cluster Upgrade

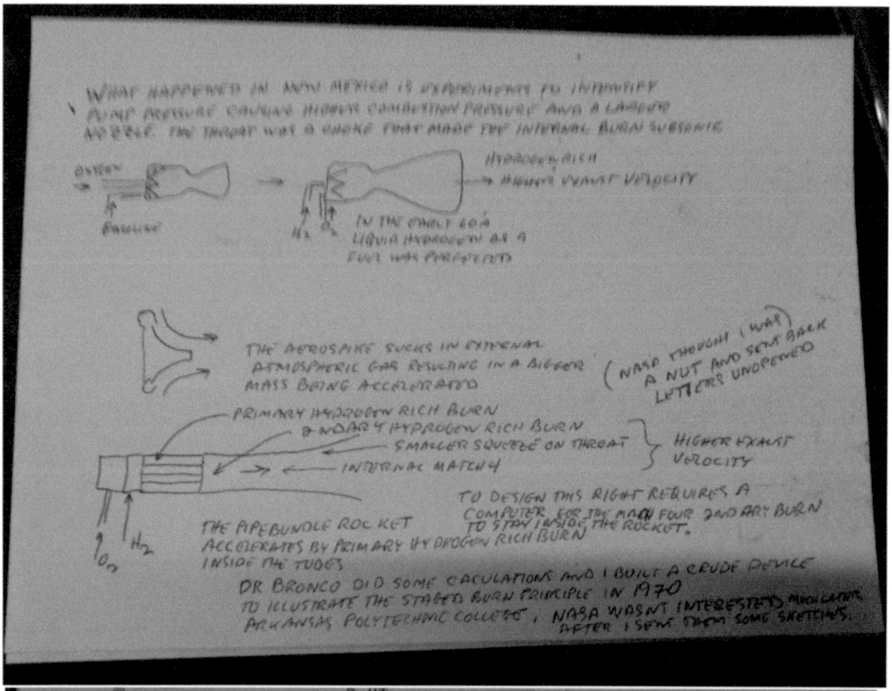

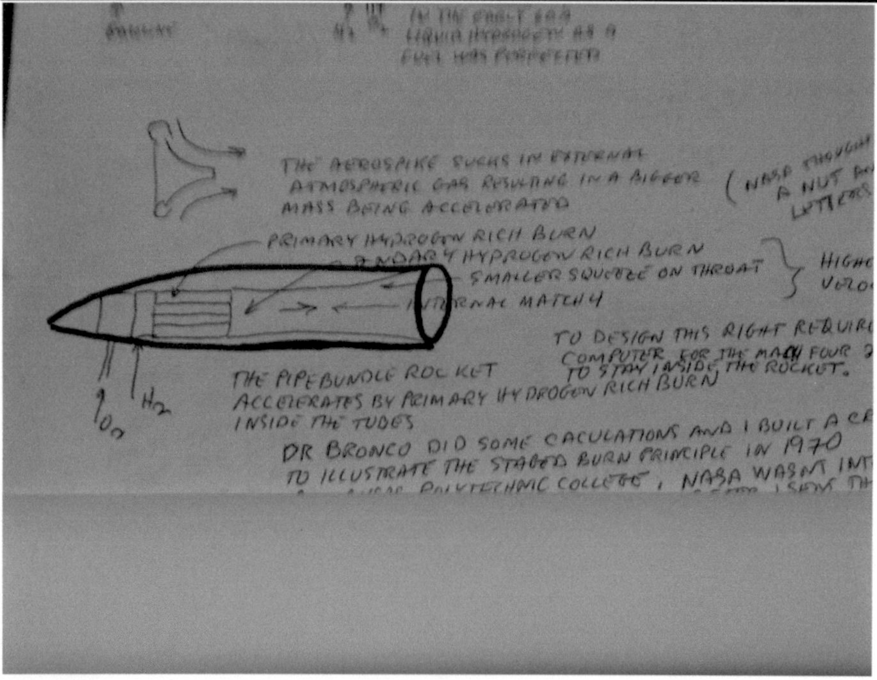

Document Scan - Hypersonic Jet

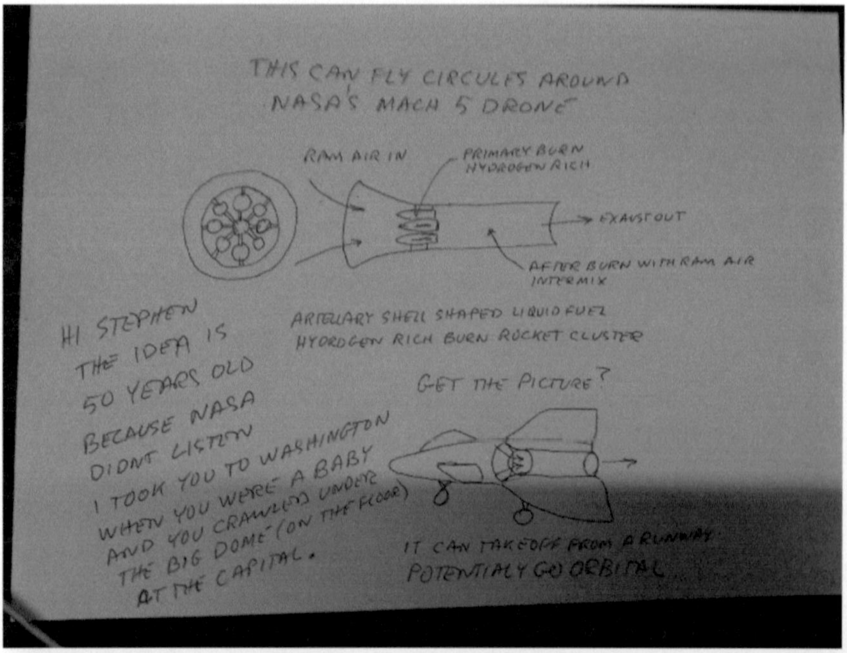

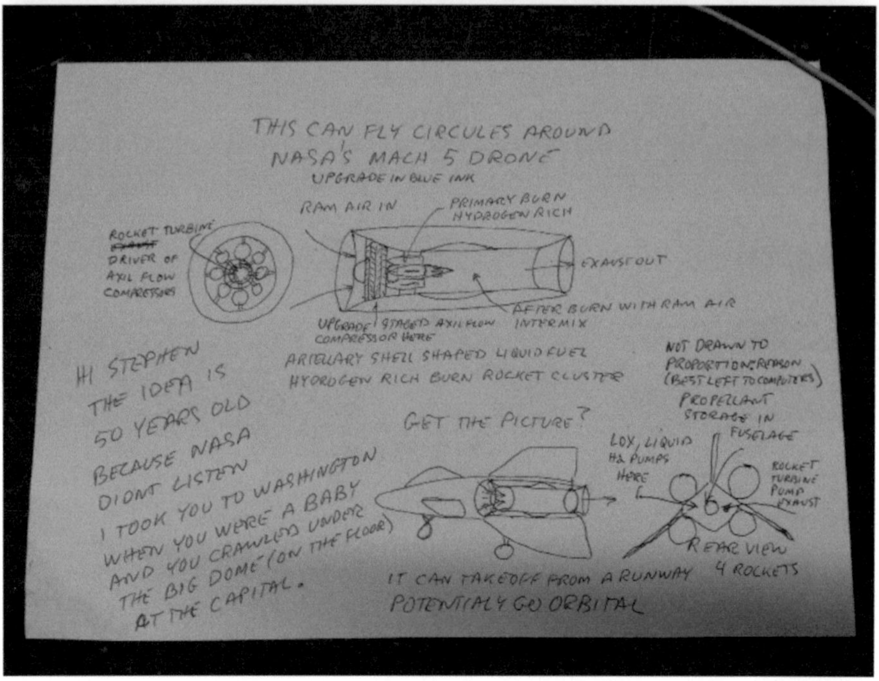

Document Scan - Cyclotron Rocket & Bouyant Blimp

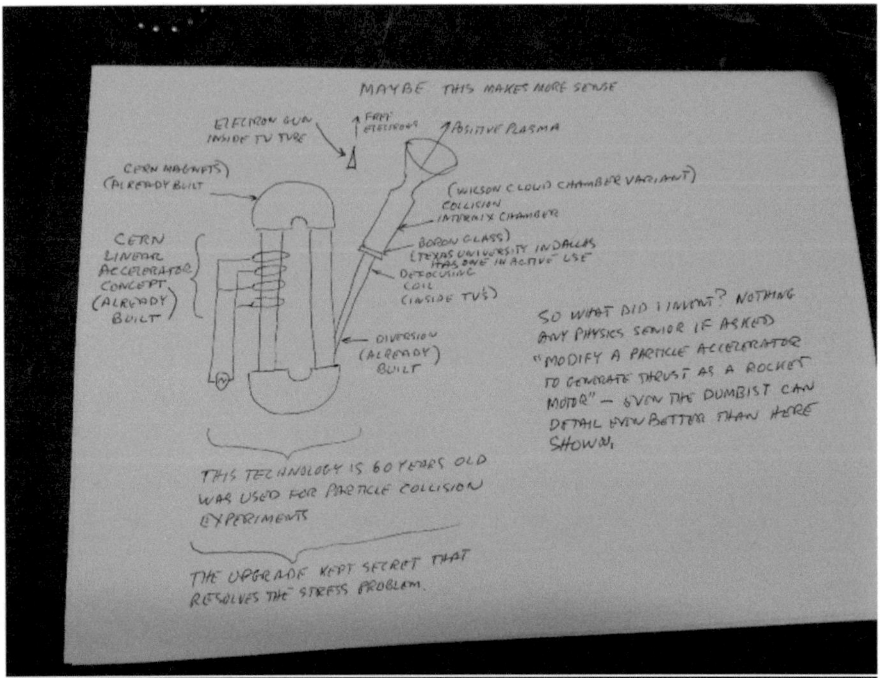

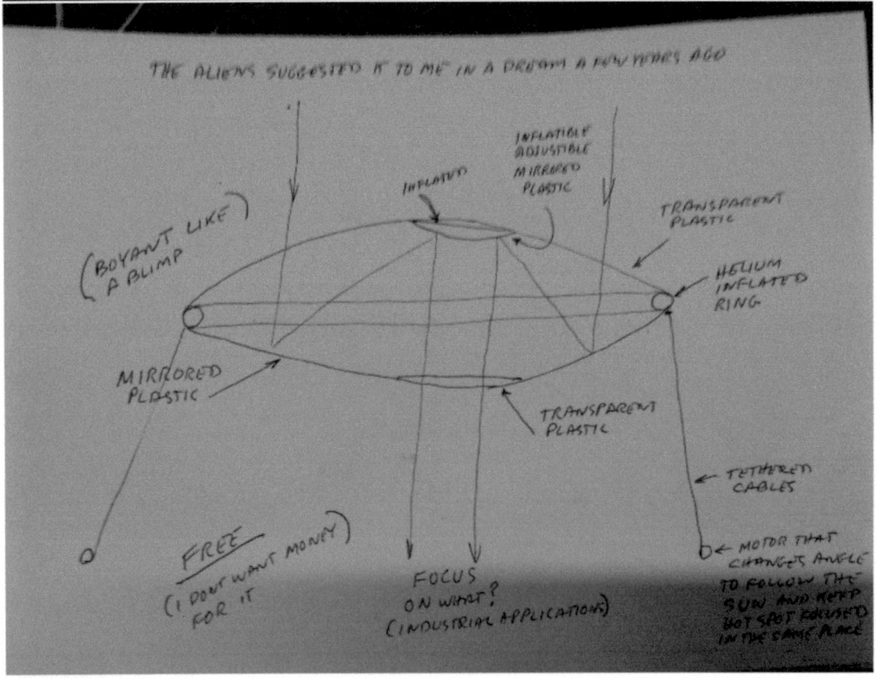

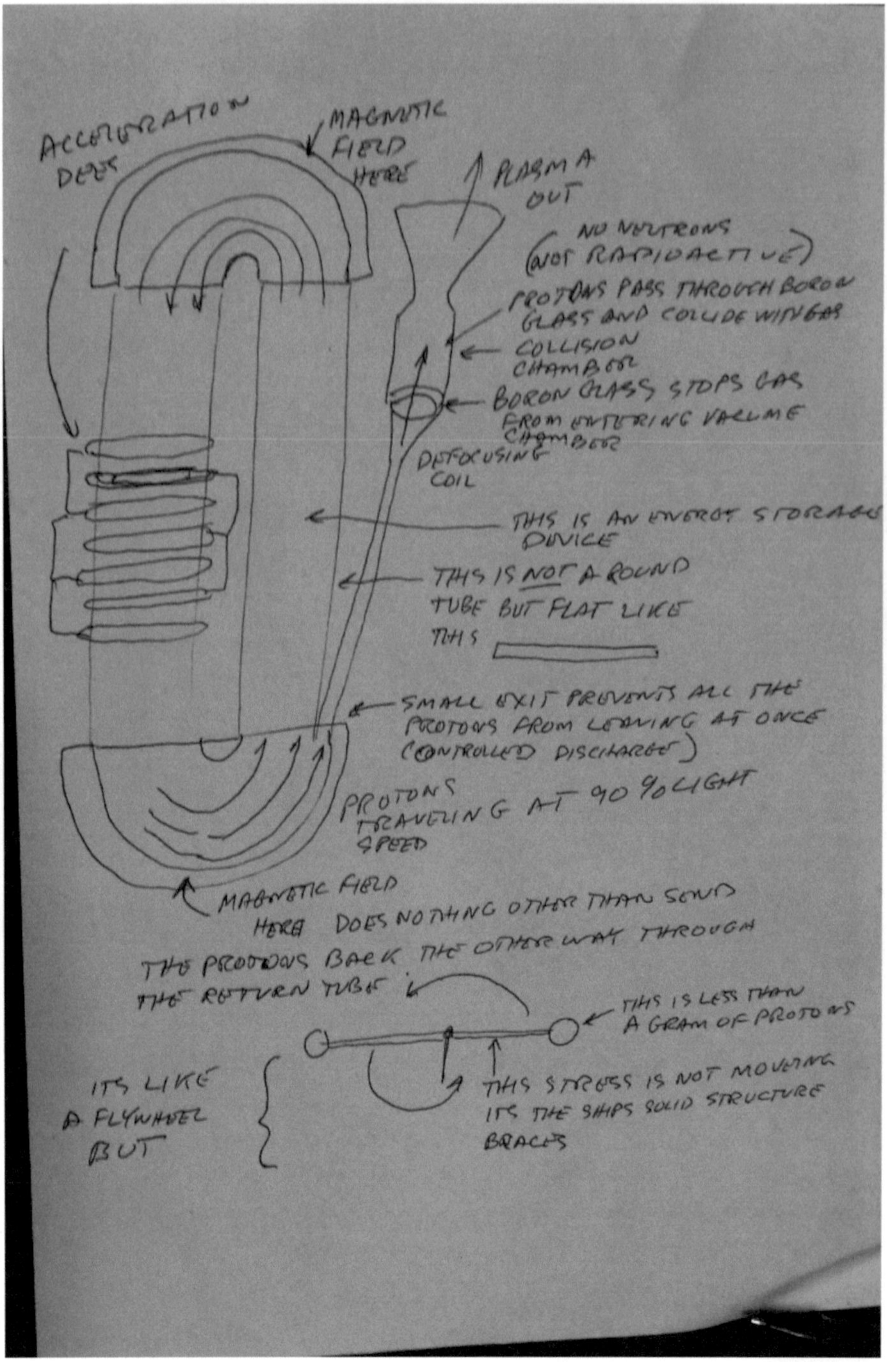
ACCELERATION DEES
MAGNETIC FIELD HERE
PLASMA OUT
NO NEUTRONS (NOT RADIOACTIVE)
PROTONS PASS THROUGH BORON GLASS AND COLLIDE WITH GAS
COLLISION CHAMBER
BORON GLASS STOPS GAS FROM ENTERING VACUME CHAMBER
DEFOCUSING COIL
THIS IS AN ENERGY STORAGE DEVICE
THIS IS NOT A ROUND TUBE BUT FLAT LIKE THIS
SMALL EXIT PREVENTS ALL THE PROTONS FROM LEAVING AT ONCE (CONTROLLED DISCHARGE)
PROTONS TRAVELING AT 90 % LIGHT SPEED
MAGNETIC FIELD HERE DOES NOTHING OTHER THAN SEND THE PROTONS BACK THE OTHER WAY THROUGH THE RETURN TUBE
THIS IS LESS THAN A GRAM OF PROTONS
ITS LIKE A FLYWHEEL BUT
THIS STRESS IS NOT MOVING ITS THE SHIPS SOLID STRUCTURE BRACES

Document Scan - Particle Accelerator Rocket (Why Two?)

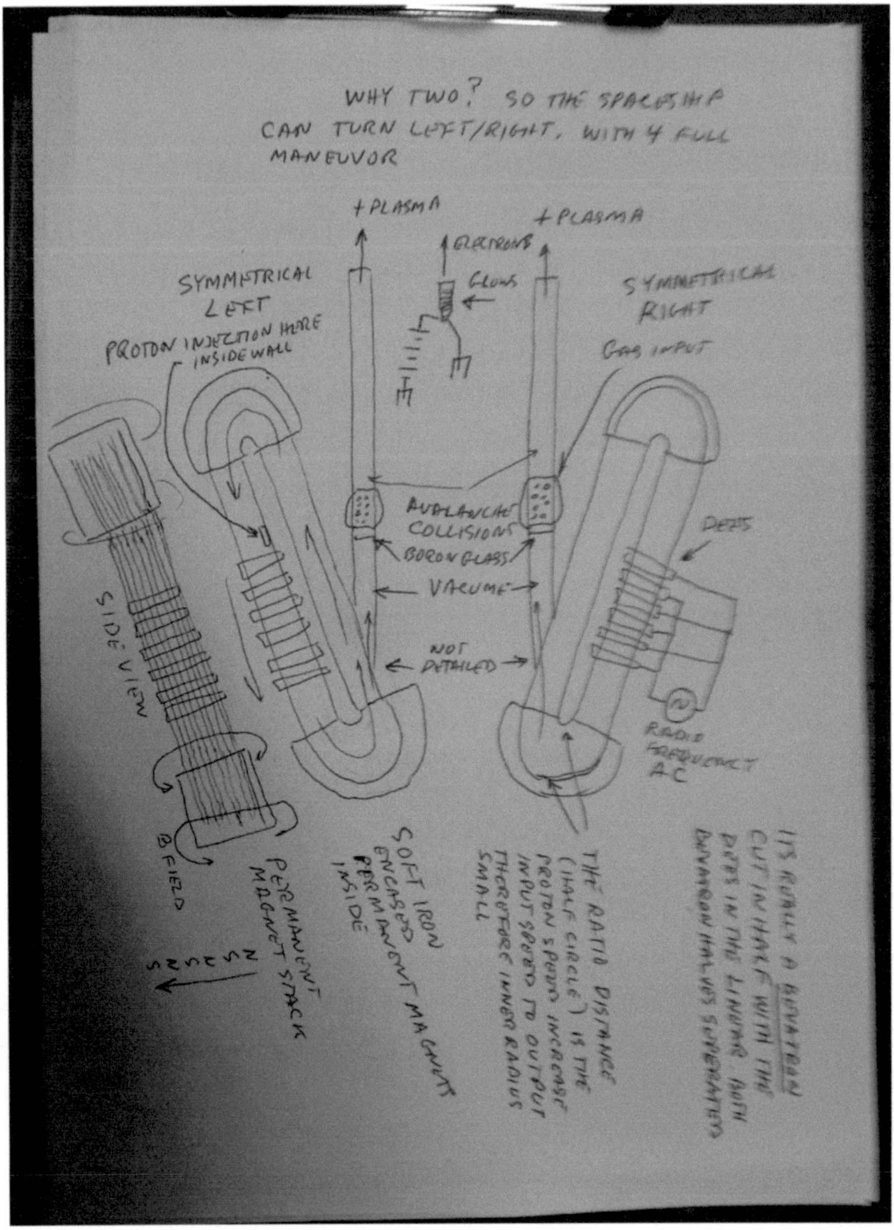

Document Scan - Particle Accelerator Rocket

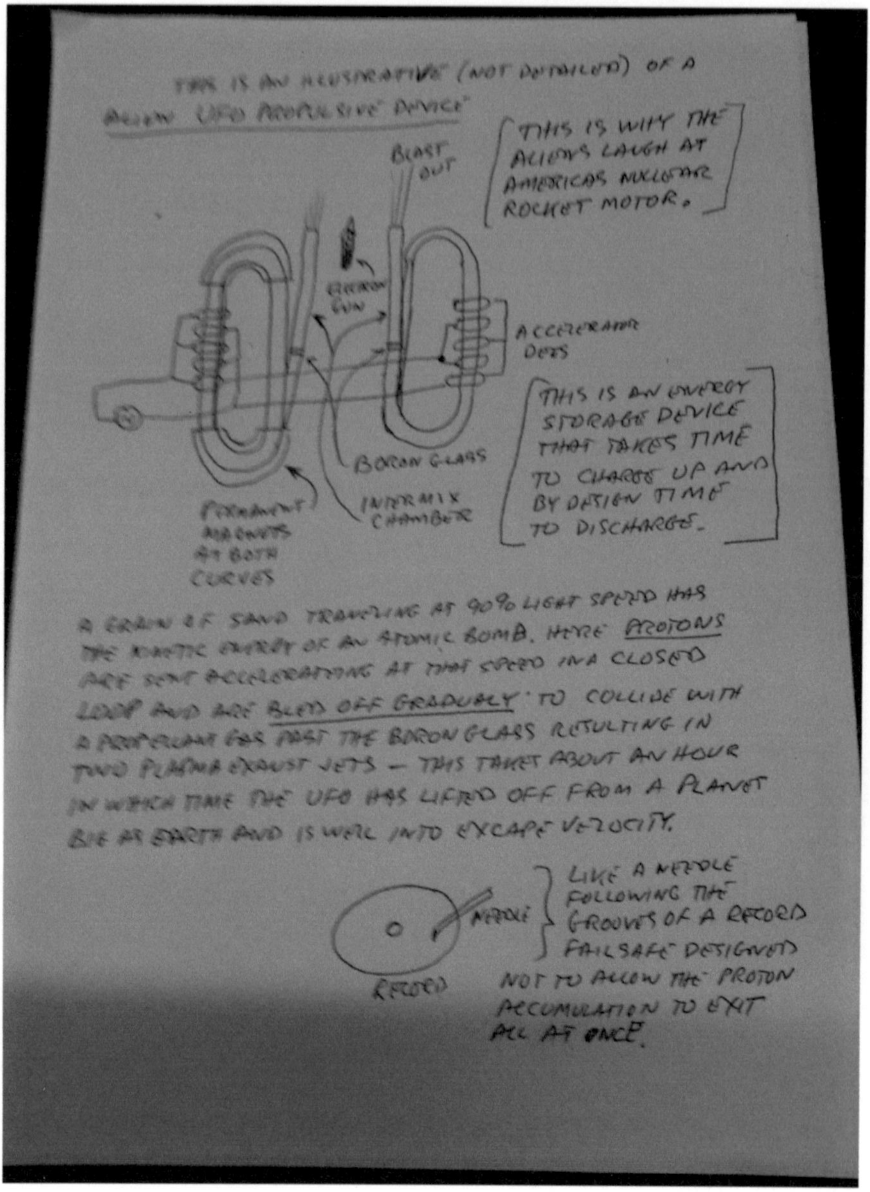

THIS IS AN ILLUSTRATIVE (NOT DETAILED) OF A
ALIEN UFO PROPULSIVE DEVICE

A GRAIN OF SAND TRAVELING AT 90% LIGHT SPEED HAS THE KINETIC ENERGY OF AN ATOMIC BOMB. HERE PROTONS ARE SENT ACCELERATING AT THAT SPEED IN A CLOSED LOOP AND ARE BLED OFF GRADUALY TO COLLIDE WITH A PROPELLANT GAS PAST THE BORON GLASS RESULTING IN TWO PLASMA EXHAUST JETS — THIS TAKES ABOUT AN HOUR IN WHICH TIME THE UFO HAS LIFTED OFF FROM A PLANET BIG AS EARTH AND IS WELL INTO ESCAPE VELOCITY.

LIKE A NEEDLE FOLLOWING THE GROOVES OF A RECORD
FAILSAFE DESIGNED NOT TO ALLOW THE PROTON ACCUMULATION TO EXIT ALL AT ONCE.

Document Scan - Particle Accelerator Rocket (Storage Device)

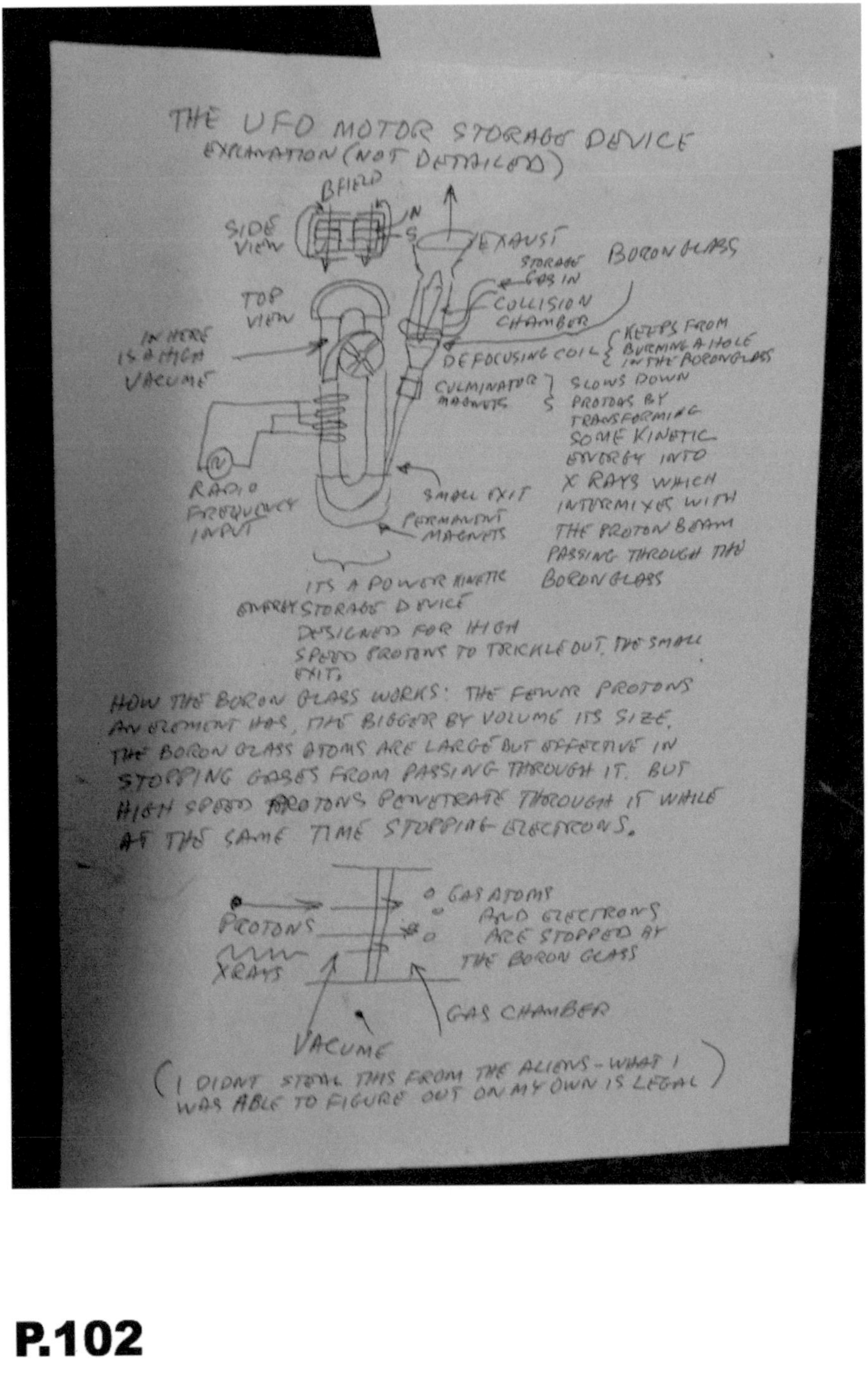
THE UFO MOTOR STORAGE DEVICE
EXPLANATION (NOT DETAILED)

HOW THE BORON GLASS WORKS: THE FEWER PROTONS AN ELEMENT HAS, THE BIGGER BY VOLUME ITS SIZE. THE BORON GLASS ATOMS ARE LARGE BUT EFFECTIVE IN STOPPING GASES FROM PASSING THROUGH IT. BUT HIGH SPEED PROTONS PENETRATE THROUGH IT WHILE AT THE SAME TIME STOPPING ELECTRONS.

(I DIDNT STEAL THIS FROM THE ALIENS - WHAT I WAS ABLE TO FIGURE OUT ON MY OWN IS LEGAL)

Document Scan - UFO Motor & Toroidal Donut

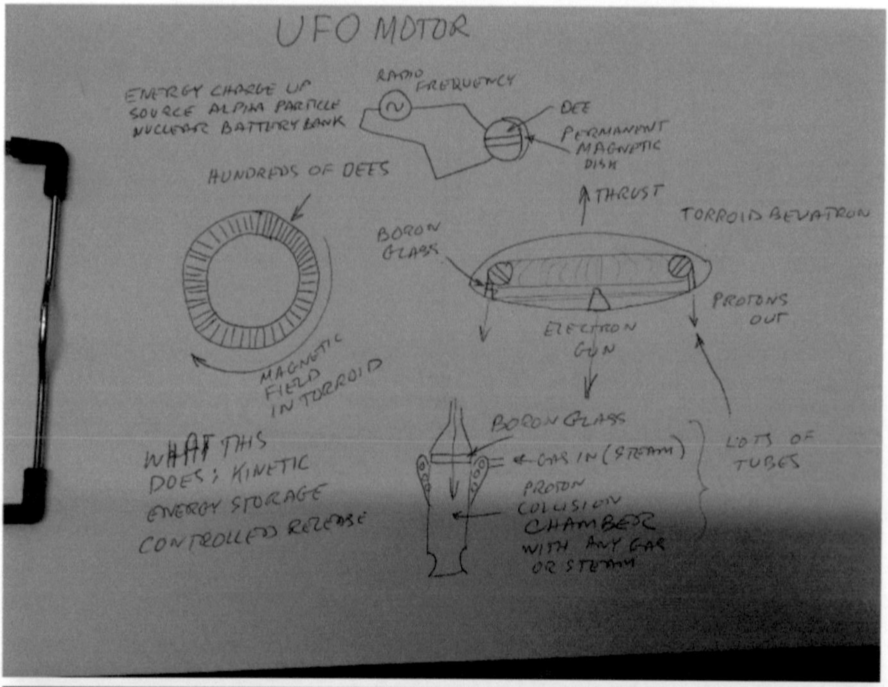

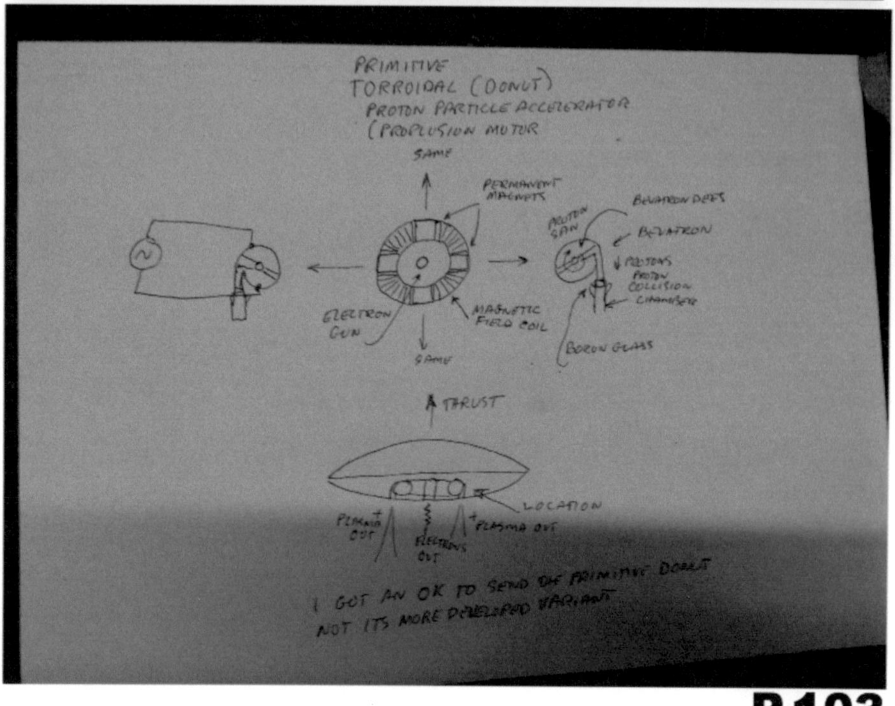

Document Scan - Toroid Bevatron

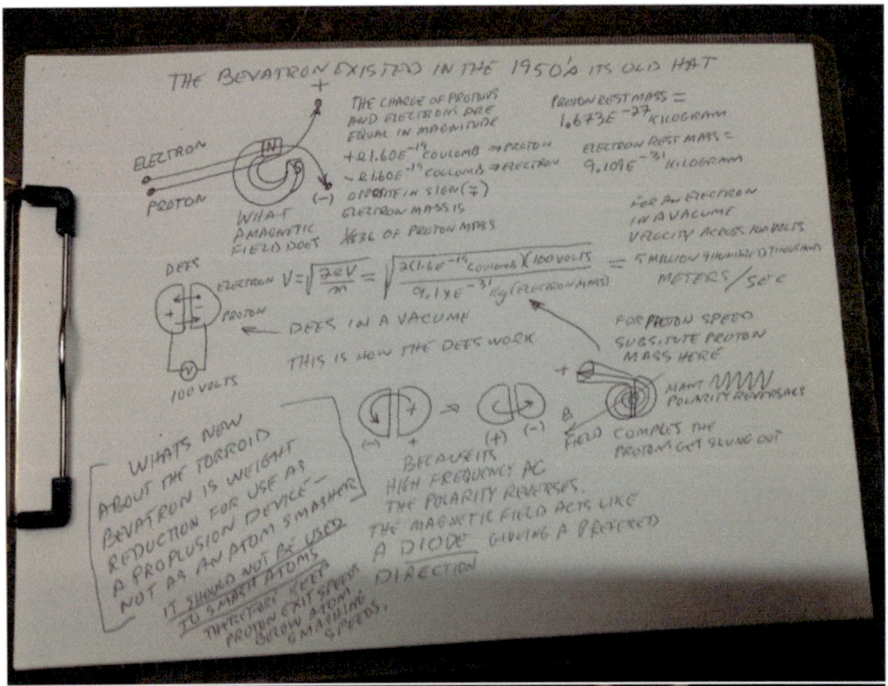

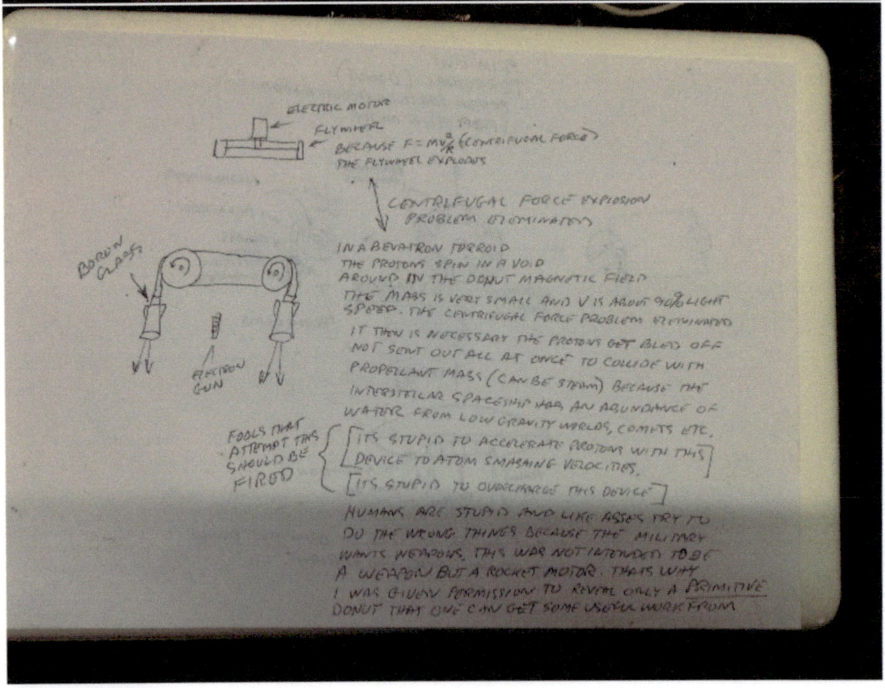

Document Scan - Design Progression & Fleet Motor

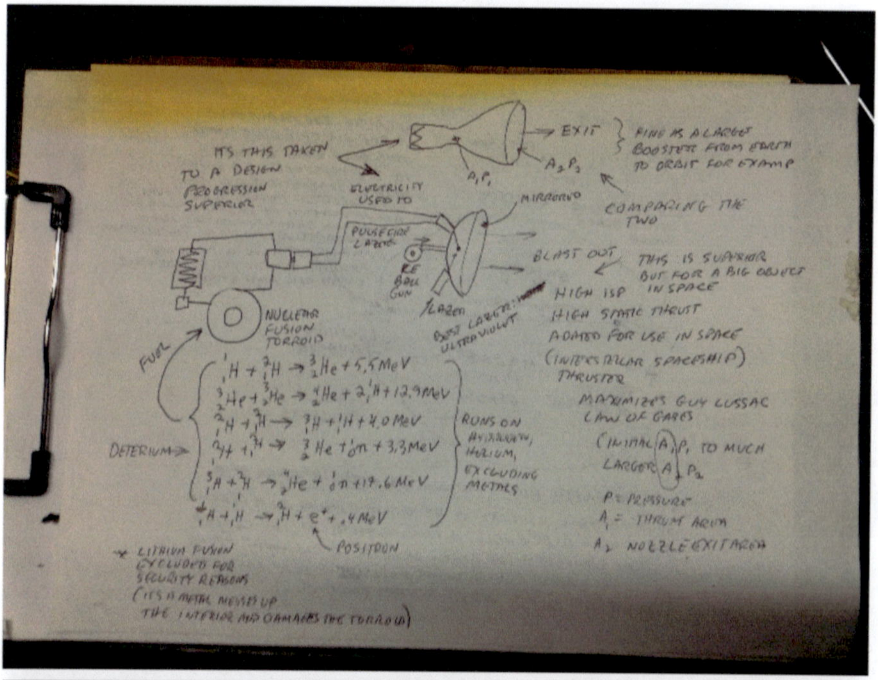

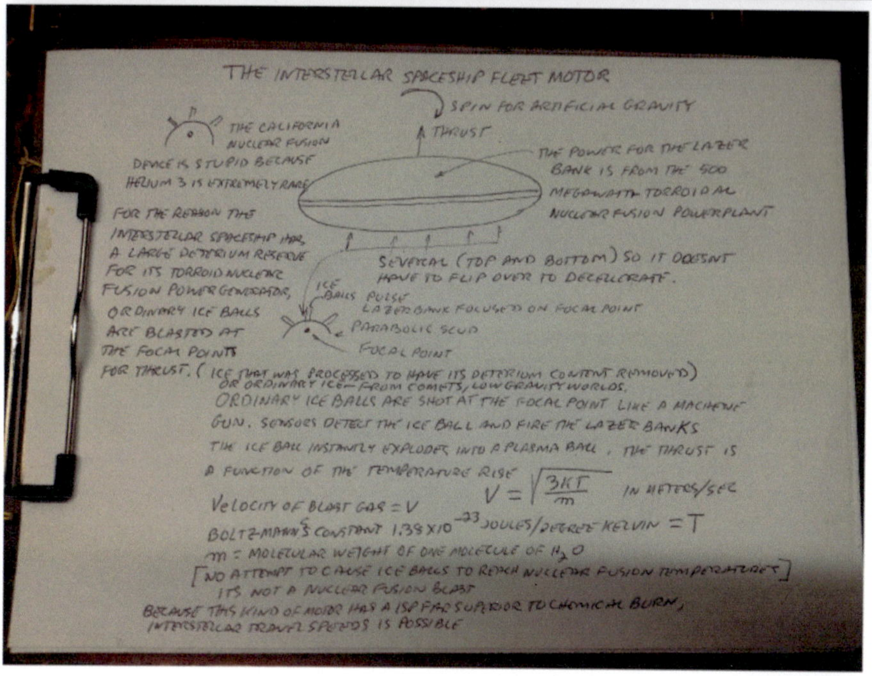

Document Scan - Toroid Bevatron

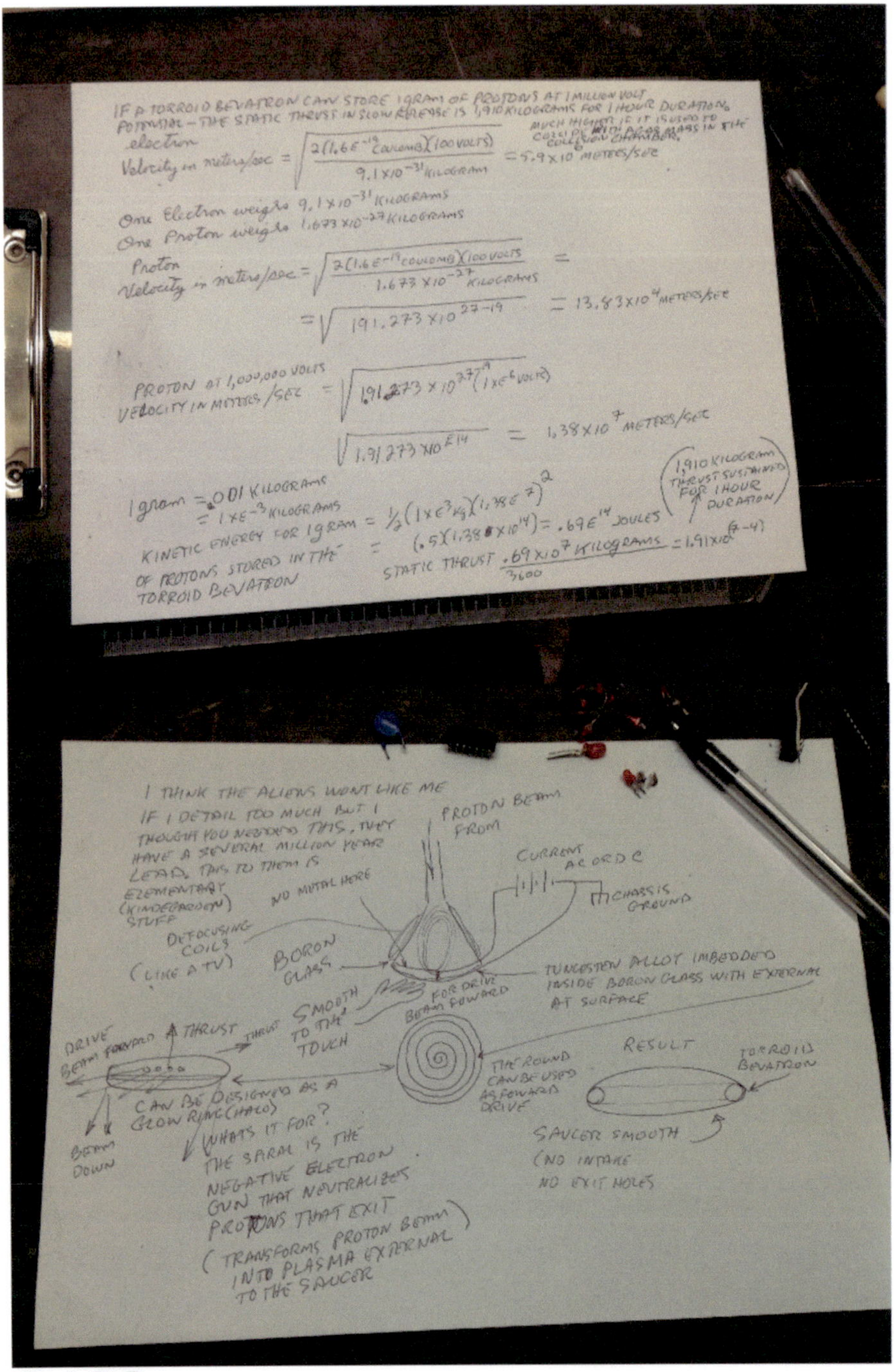

Document Scan - Motor Matrix

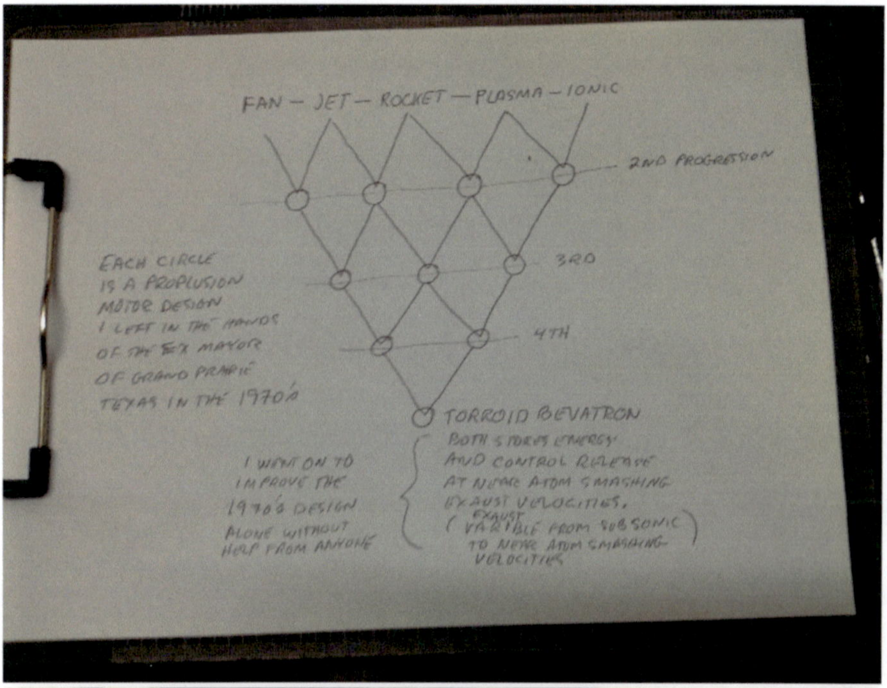

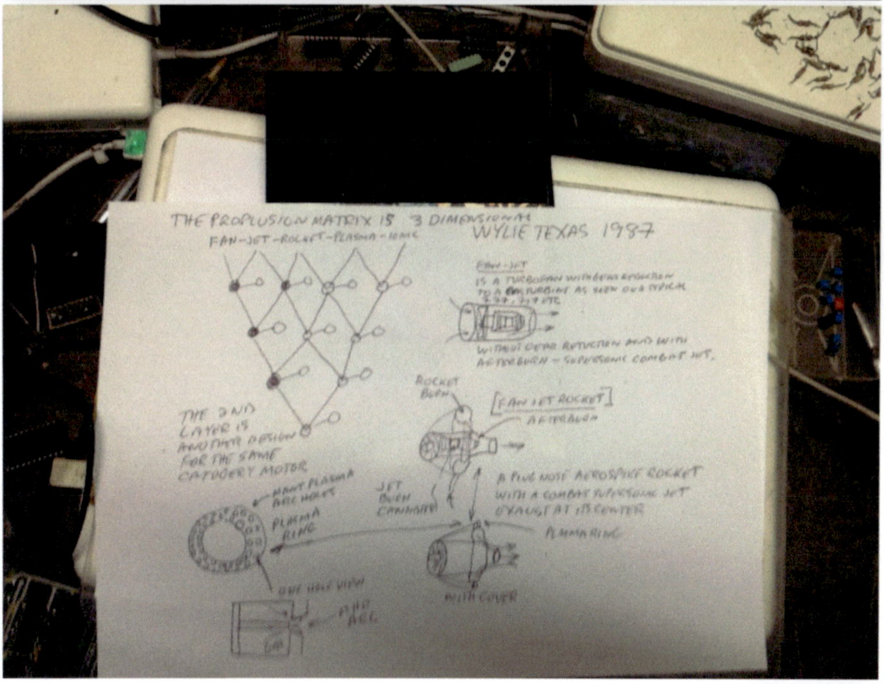

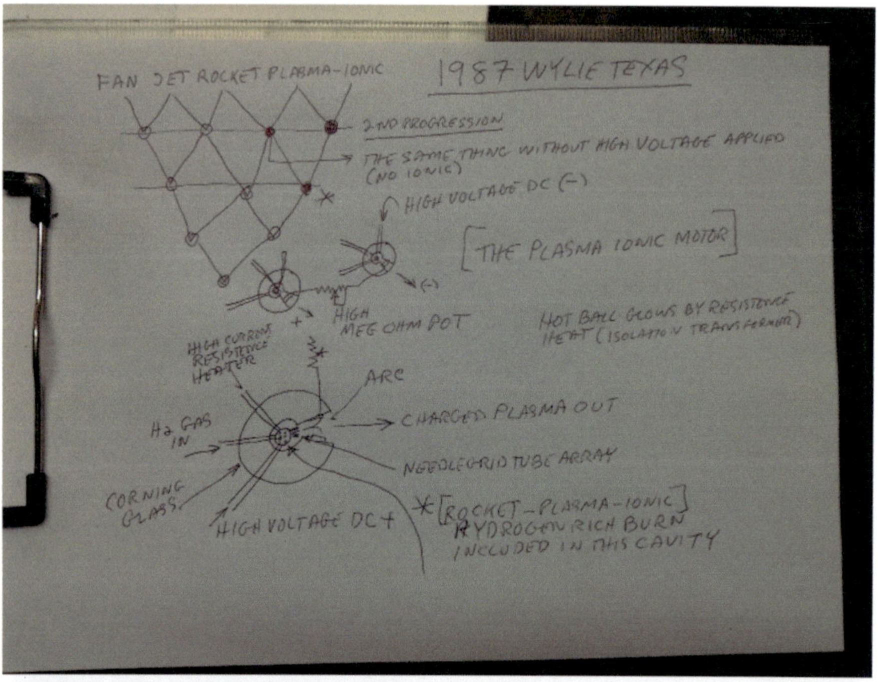
FAN JET ROCKET PLASMA-IONIC
1987 WYLIE TEXAS
2ND PROGRESSION
THE SAME THING WITHOUT HIGH VOLTAGE APPLIED (NO IONIC)
HIGH VOLTAGE DC (-)
[THE PLASMA IONIC MOTOR]
HIGH MEG OHM POT
HOT BALL GLOWS BY RESISTENCE HEAT (ISOLATION TRANSFORMER)
HIGH CURRENT RESISTENCE HEATER
ARC
CHARGED PLASMA OUT
H2 GAS IN
NEEDLEGRID TUBE ARRAY
CORNING GLASS
HIGH VOLTAGE DC+
*[ROCKET-PLASMA-IONIC]
HYDROGEN RICH BURN
INCLUDED IN THIS CAVITY

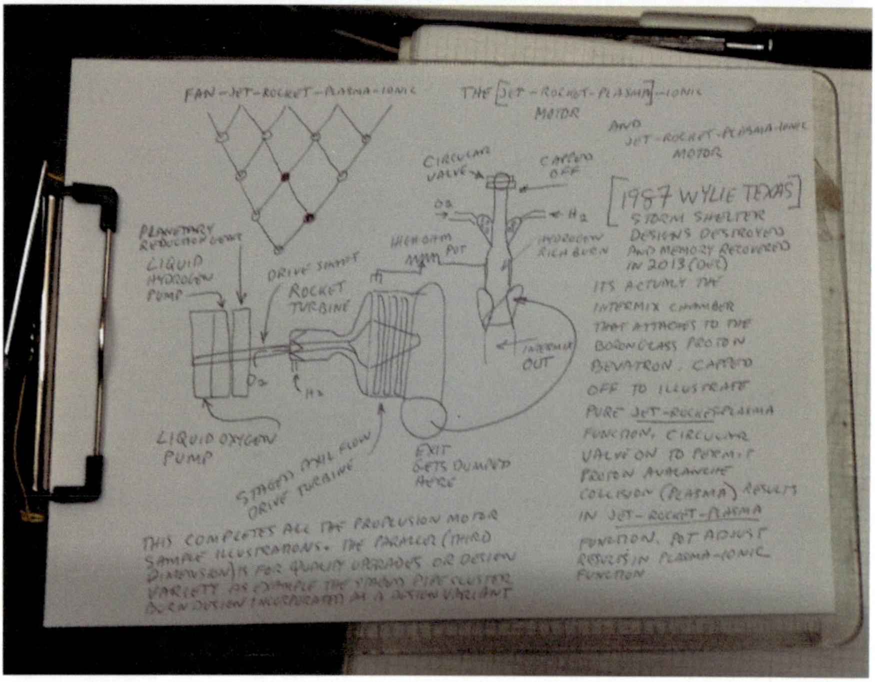
FAN-JET-ROCKET-PLASMA-IONIC
THE [JET-ROCKET-PLASMA]-IONIC MOTOR
AND
JET-ROCKET-PLASMA-IONIC MOTOR
CIRCULAR VALVE
CAPPED OFF
[1987 WYLIE TEXAS]
STORM SHELTER DESIGNS DESTROYED AND MEMORY RECOVERED IN 2013 (OCT)
PLANETARY REDUCTION GEAR
LIQUID HYDROGEN PUMP
DRIVE SHAFT
ROCKET TURBINE
HIGH OHM POT
HYDROGEN RICH BURN
INTERMIX OUT
ITS ACTUALLY THE INTERMIX CHAMBER THAT ATTACHES TO THE BORONGLASS PROTON BEVATRON. CAPPED OFF TO ILLUSTRATE PURE JET-ROCKET-PLASMA FUNCTION, CIRCULAR VALVE ON TO PERMIT PROTON AVALANCHE COLLISION (PLASMA) RESULTS IN JET-ROCKET-PLASMA FUNCTION. POT ADJUST RESULTS IN PLASMA-IONIC FUNCTION
LIQUID OXYGEN PUMP
STAGED AXIL FLOW DRIVE TURBINE
EXIT GETS DUMPED HERE
THIS COMPLETES ALL THE PROPULSION MOTOR SAMPLE ILLUSTRATIONS. THE PARALLEL (THIRD DIMENSION) IS FOR QUALITY UPGRADES OR DESIGN VARIETY AS EXAMPLE THE STAGED PIPE CLUSTER BURN DESIGN INCORPORATED AS A DESIGN VARIANT

Document Scan - Cigar UFO

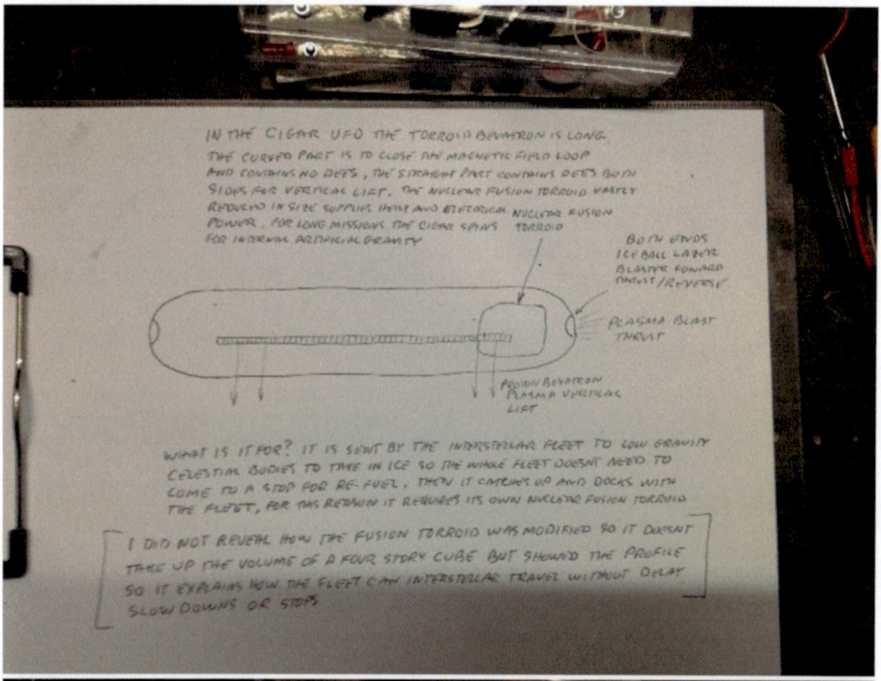

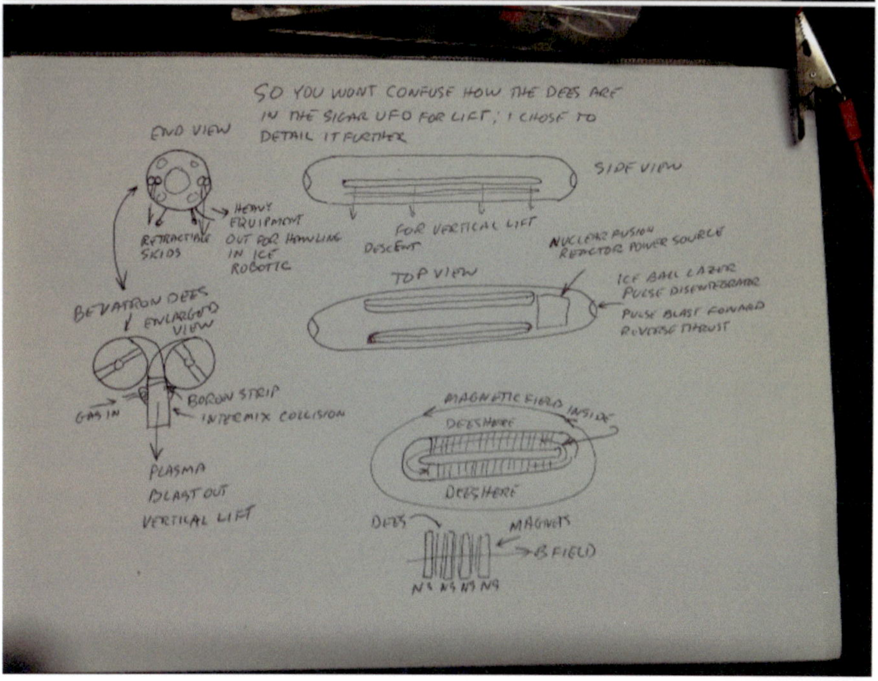

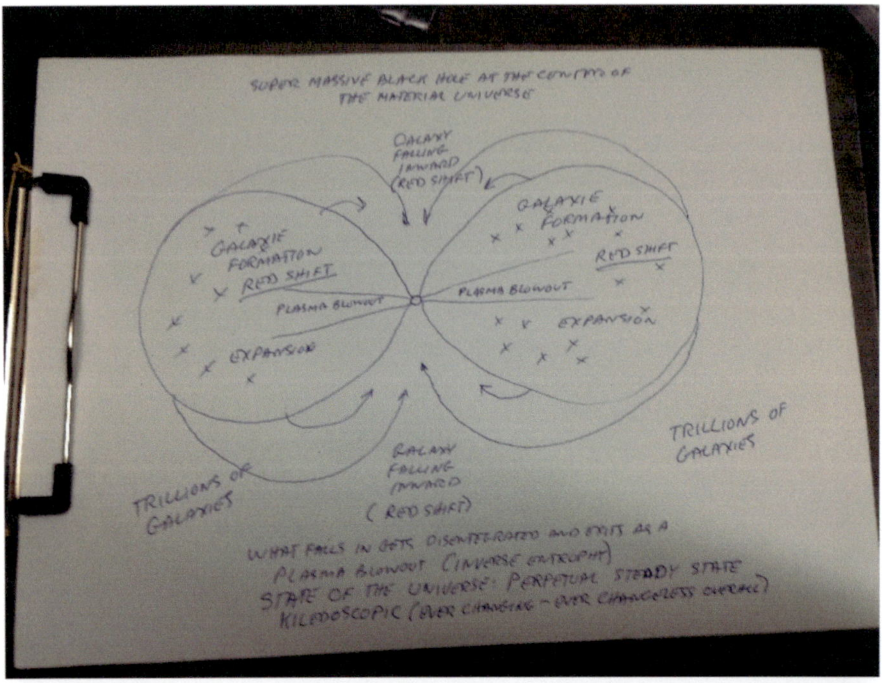
SUPER MASSIVE BLACK HOLE AT THE CENTER OF THE MATERIAL UNIVERSE
GALAXY FALLING INWARD (RED SHIFT)
GALAXIE FORMATION
RED SHIFT
PLASMA BLOWOUT
EXPANSION
GALAXIE FORMATION
RED SHIFT
PLASMA BLOWOUT
EXPANSION
TRILLIONS OF GALAXIES
GALAXY FALLING INWARD (RED SHIFT)
TRILLIONS OF GALAXIES
WHAT FALLS IN GETS DISINTEGRATED AND EXITS AS A PLASMA BLOWOUT (INVERSE ENTROPHY)
STATE OF THE UNIVERSE: PERPETUAL STEADY STATE KILEDOSCOPIC (EVER CHANGING - EVER CHANGELESS OVERALL)

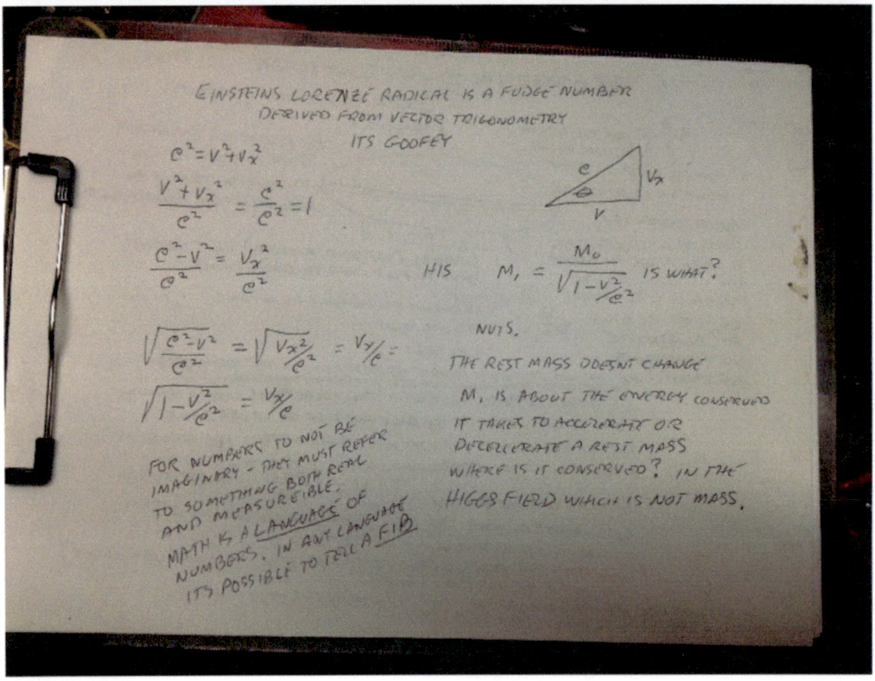
EINSTEINS LORENZ RADICAL IS A FUDGE NUMBER DERIVED FROM VECTOR TRIGONOMETRY
ITS GOOFEY
HIS
IS WHAT?
NUTS.
THE REST MASS DOESNT CHANGE
M, IS ABOUT THE ENERGY CONSERVED IT TAKES TO ACCELERATE OR DECELERATE A REST MASS
WHERE IS IT CONSERVED? IN THE HIGGS FIELD WHICH IS NOT MASS.
FOR NUMBERS TO NOT BE IMAGINARY - THEY MUST REFER TO SOMETHING BOTH REAL AND MEASUREIBLE.
MATH IS A LANGUAGE OF NUMBERS. IN ANY LANGUAGE ITS POSSIBLE TO TELL A FIB

Document Scan - 10,000 BC & Saucer Dream

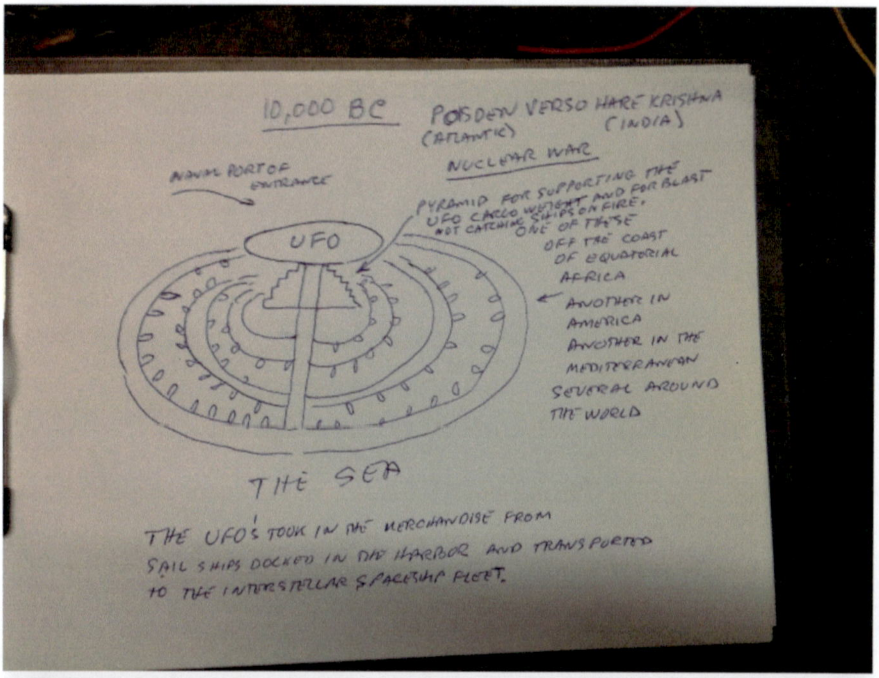

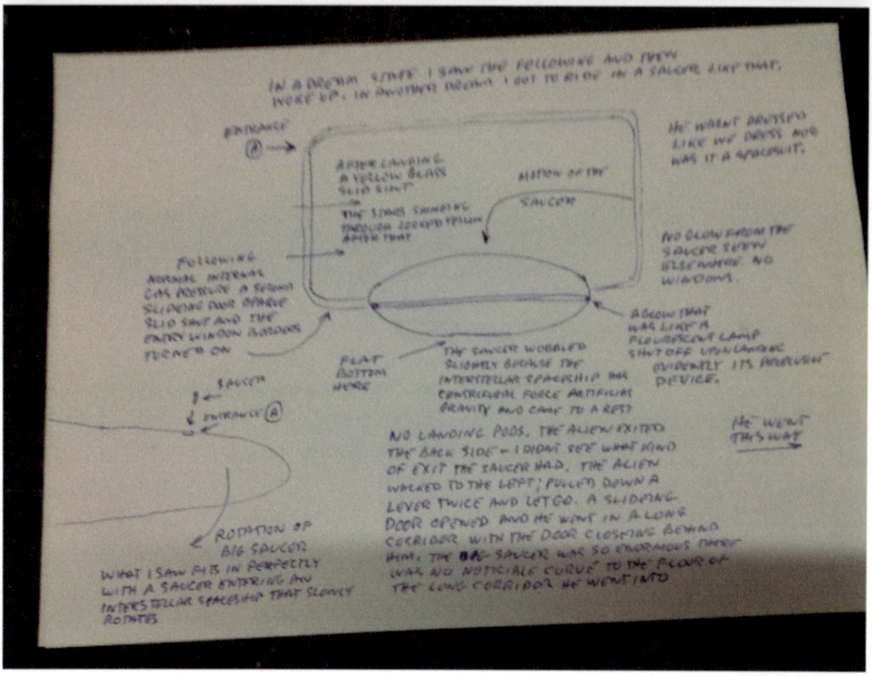

Document Scan - Propulsion Progression

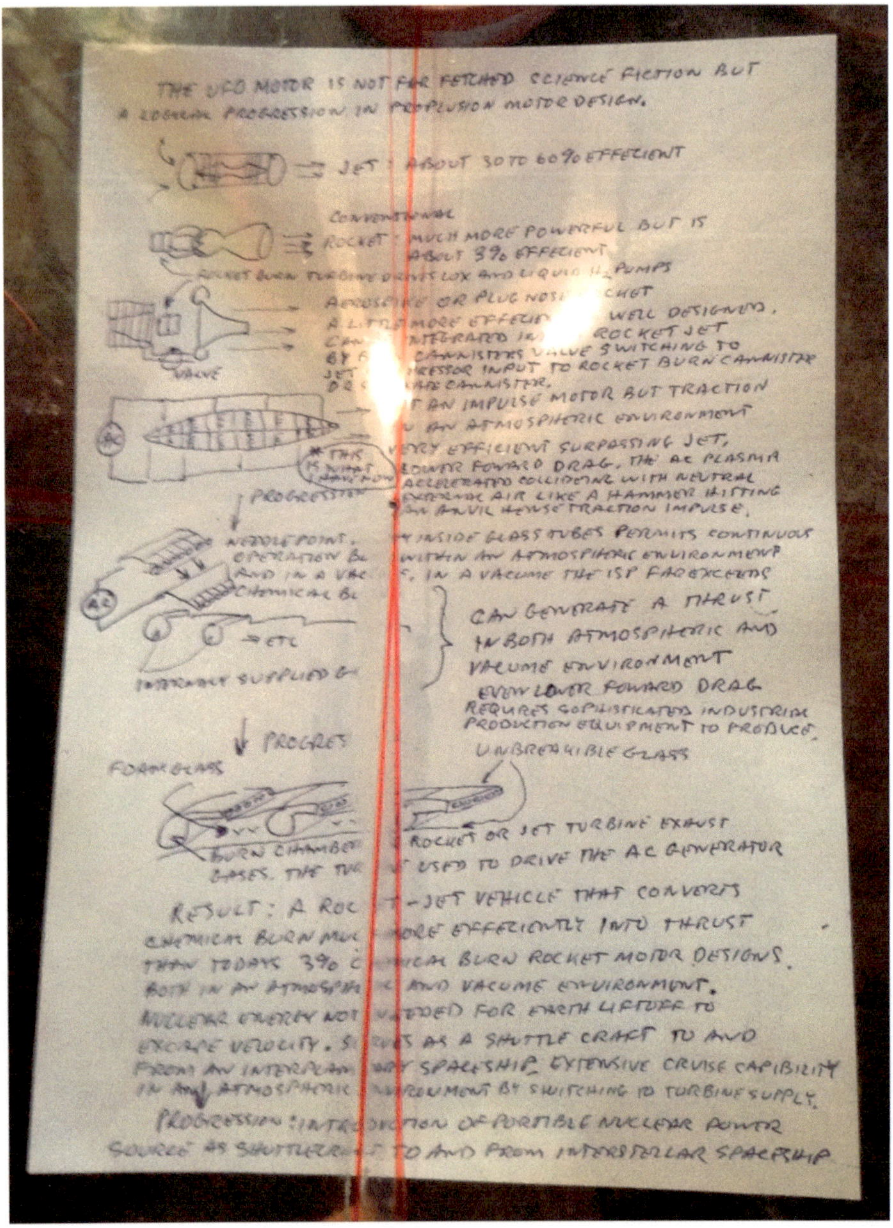
THE UFO MOTOR IS NOT FAR FETCHED SCIENCE FICTION BUT
A LOGICAL PROGRESSION IN PROPULSION MOTOR DESIGN.

JET : ABOUT 30 TO 60% EFFECIENT

CONVENTIONAL
ROCKET : MUCH MORE POWERFUL BUT IS
ABOUT 3% EFFECIENT
ROCKET BURN TURBINE DRIVES LOX AND LIQUID H_2 PUMPS

AEROSPIKE OR PLUG NOSE ROCKET
A LITTLE MORE EFFECIENT [illegible] WELL DESIGNED,
CAN [illegible] INTEGRATED IN [illegible] ROCKET JET
BY [illegible] CANNISTERS VALVE SWITCHING TO
JET [illegible] PRESSOR INPUT TO ROCKET BURN CANNISTER
OR [illegible] CANNISTER.
VALVE

[illegible] AN IMPULSE MOTOR BUT TRACTION
[illegible] AN ATMOSPHERIC ENVIRONMENT
VERY EFFICIENT SURPASSING JET,
LOWER FOWARD DRAG. THE AC PLASMA
ACCELERATED COLLIDING WITH NEUTRAL
EXTERNAL AIR LIKE A HAMMER HITTING
AN ANVIL HENSE TRACTION IMPULSE.
* THIS IS WHAT I HAVE NOW
PROGRESSION

NEEDLE POINT. [illegible] INSIDE GLASS TUBES PERMITS CONTINUOUS
OPERATION B[illegible] WITHIN AN ATMOSPHERIC ENVIRONMENT
AND IN A VAC[illegible], IN A VACUME THE ISP FAR EXCEEDS
CHEMICAL B[illegible]
→ ETC
INTERNALLY SUPPLIED G[illegible]

CAN GENERATE A THRUST
IN BOTH ATMOSPHERIC AND
VACUME ENVIRONMENT
EVEN LOWER FOWARD DRAG
REQUIRES SOPHISTICATED INDUSTRIAL
PRODUCTION EQUIPMENT TO PRODUCE.

PROGRES[illegible]
FOAM GLASS
UNBREAKIBLE GLASS

BURN CHAMBE[illegible] ROCKET OR JET TURBINE EXAUST
GASES. THE TUR[illegible] USED TO DRIVE THE AC GENERATOR

RESULT: A ROC[illegible]-JET VEHICLE THAT CONVERTS
CHEMICAL BURN MU[illegible] MORE EFFECIENTLY INTO THRUST
THAN TODAYS 3% C[illegible]MICAL BURN ROCKET MOTOR DESIGNS.
BOTH IN AN ATMOSPH[illegible]IC AND VACUME ENVIRONMENT.
NUCLEAR ENERGY NOT [illegible]EEDED FOR EARTH LIFTOFF TO
ESCAPE VELOCITY. S[illegible]ES AS A SHUTTLE CRAFT TO AND
FROM AN INTERPLAN[illegible]ARY SPACESHIP. EXTENSIVE CRUISE CAPIBILITY
IN AN ATMOSPHERIC [illegible]NVIRONMENT BY SWITCHING TO TURBINE SUPPLY.
PROGRESSION: INTRODUCTION OF PORTIBLE NUCLEAR POWER
SOURCE AS SHUTTLECRA[illegible]T TO AND FROM INTERSTELLAR SPACESHIP

Document Scan - Nuclear Fusion Toroid Reactor

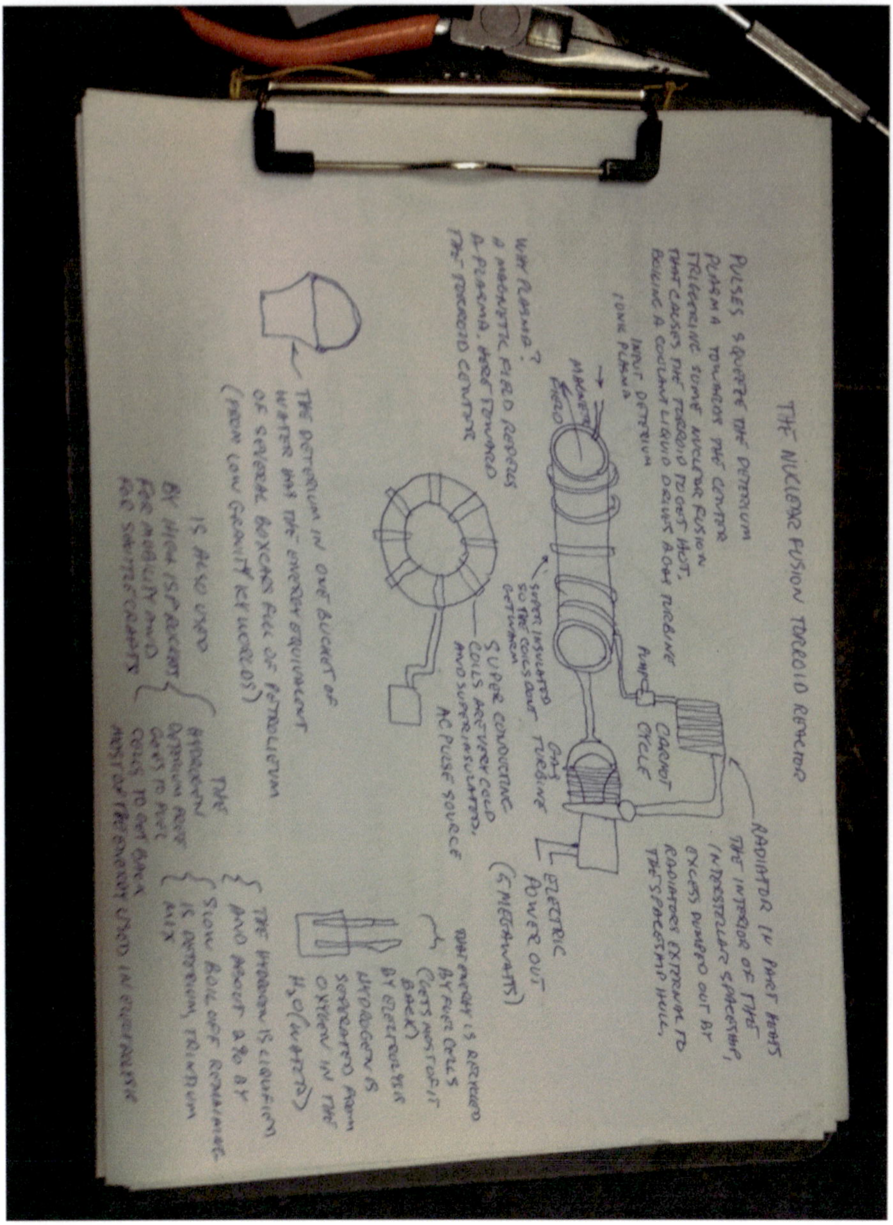

WHAT I HAD IN THE TORNADO SHELTER WAS A DC IONIC BREEZE COMING FROM ONE SINGLE NEEDLE POINT I COULD FEEL WITH MY HAND OVER A YARD AWAY; AND I COULD MAKE THE IONS POSITIVE OR NEGATIVE BY FLIPPING OVER THE HIGH VOLTAGE DIODE, IF WYLIE HAD LET ME FINISH MY SAUCER I WOULD HAVE OUTPACED THE USAF JETS OF THAT TIME. MY AC IONIC MOTOR COMPACTS THE DESIGN, WITH DC REQUIRES SEPERATE MOTORS AT CONSIDERABLE DISTANCE SEPERATION. AC ELIMINATES THAT. NOW THEN AS FOR THE BEER BOTTLE APPARATUS; THE BEER BOTTLES ARE HIGH VOLTAGE CAPACITORS FOR THE REASON HOME BUILT - ELECTRONIC STORES WONT SELL STUFF RESERVED FOR BUSINESS AND RESEARCH INSTITUTES THAT HAVE ALOT OF MONEY AND LAWYERS.

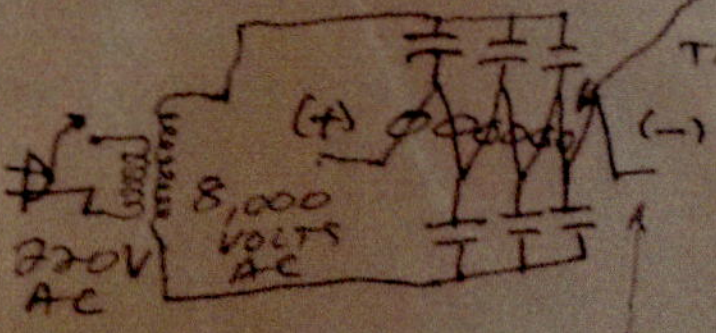

DIODES HIGH VOLTAGE FROM JUNKED TV SETS.

TAKE YOUR PICK, POSITIVE IONS OR NEGATIVE?

GROUND ONE AND GET THE OTHER OR BOTH AS TWO SEPERATE IONIC MOTORS - ONE POSITIVE THE OTHER NEGATIVE

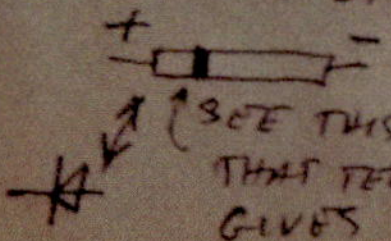

SEE THIS BAND?

THAT TELLS YOU WHICH END GIVES YOU HIGH NEGATIVE AND HIGH POSITIVE

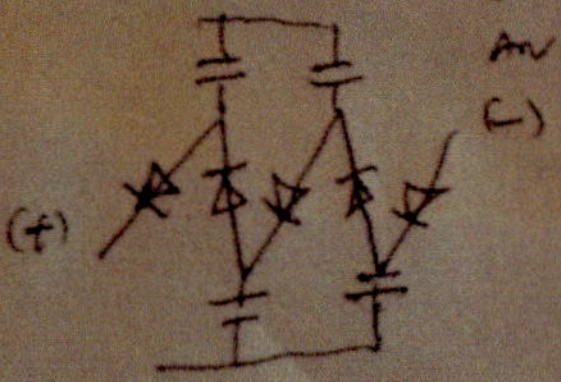

IN THE TORNADO SHELTER I WAS USEING A MUCH HIGHER FREQUENCY AND ALL MY STUFF WAS FROM JUNK. MY AC MOTOR IS WEAK BECAUSE I'M USEING JUNK - WHAT OTHERS HAVE DISCARDED FOR POWER.

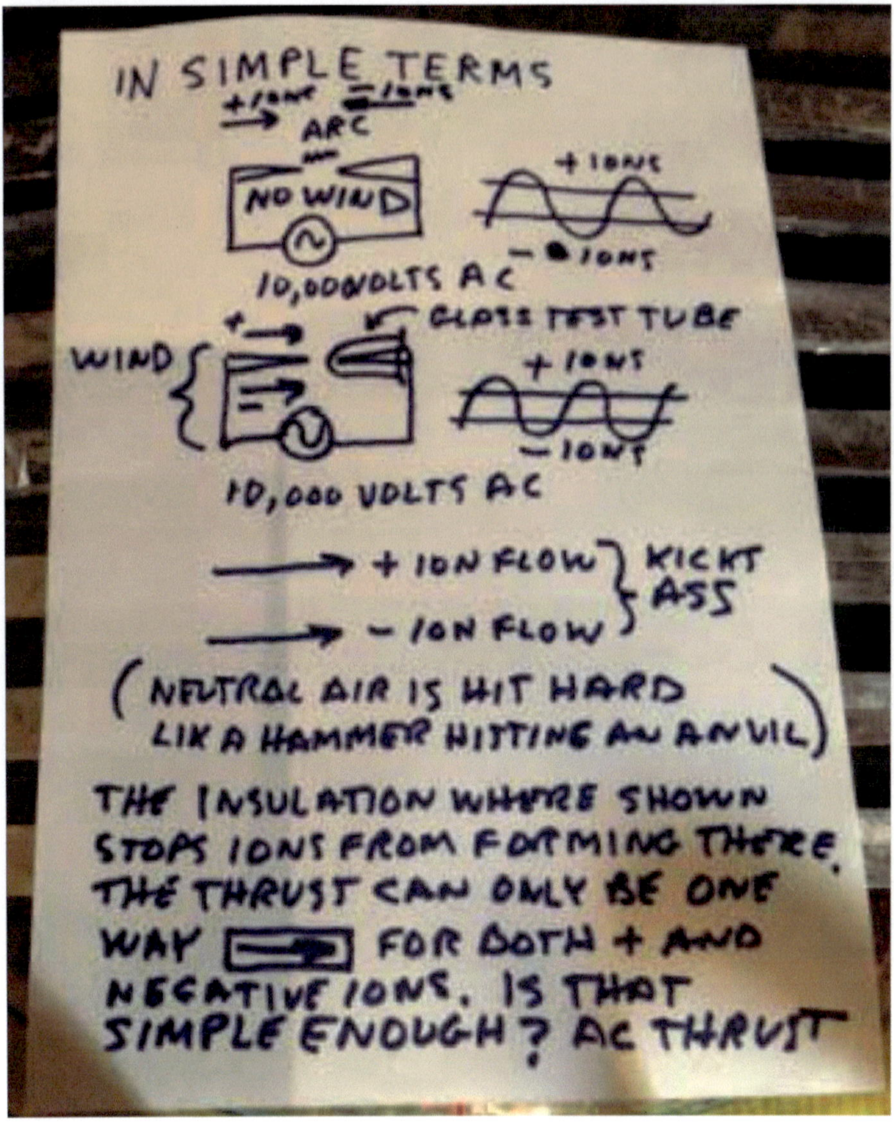

IN SIMPLE TERMS

+ ION FLOW
- ION FLOW
} KICK ASS

(NEUTRAL AIR IS HIT HARD
LIK A HAMMER HITTING AN ANVIL)

THE INSULATION WHERE SHOWN
STOPS IONS FROM FORMING THERE.
THE THRUST CAN ONLY BE ONE
WAY [→] FOR BOTH + AND
NEGATIVE IONS. IS THAT
SIMPLE ENOUGH? AC THRUST

Document Scan - Electric Field

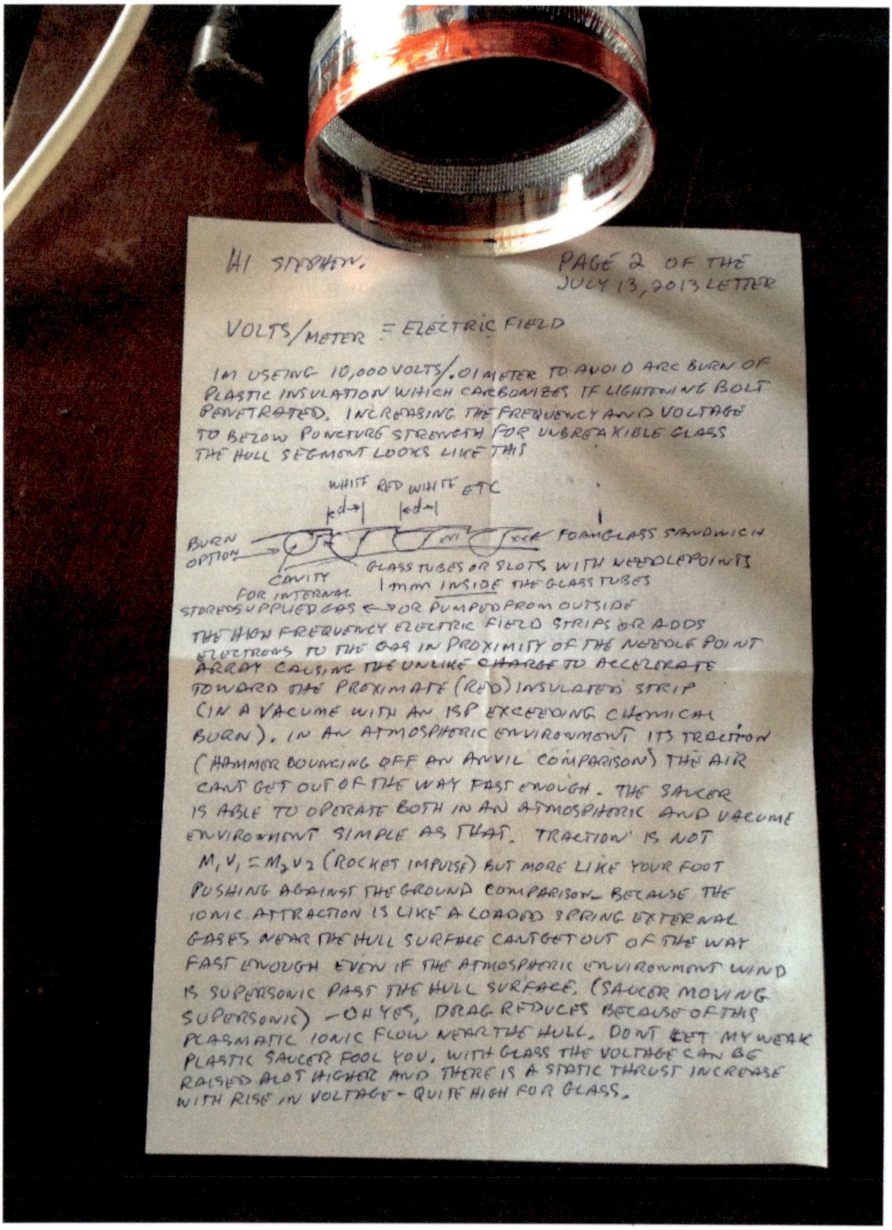

HI STEPHEN, PAGE 2 OF THE JULY 13, 2013 LETTER

VOLTS/METER = ELECTRIC FIELD

IM USEING 10,000 VOLTS/.01 METER TO AVOID ARC BURN OF PLASTIC INSULATION WHICH CARBONIZES IF LIGHTENING BOLT PENETRATED. INCREASING THE FREQUENCY AND VOLTAGE TO BELOW PUNCTURE STRENGTH FOR UNBREAKIBLE GLASS THE HULL SEGMENT LOOKS LIKE THIS

THE HIGH FREQUENCY ELECTRIC FIELD STRIPS OR ADDS ELECTRONS TO THE GAS IN PROXIMITY OF THE NEEDLE POINT ARRAY CAUSING THE UNLIKE CHARGE TO ACCELERATE TOWARD THE PROXIMATE (RED) INSULATED STRIP (IN A VACUME WITH AN ISP EXCEEDING CHEMICAL BURN). IN AN ATMOSPHERIC ENVIRONMENT ITS TRACTION (HAMMER BOUNCING OFF AN ANVIL COMPARISON) THE AIR CANT GET OUT OF THE WAY FAST ENOUGH. THE SAUCER IS ABLE TO OPERATE BOTH IN AN ATMOSPHERIC AND VACUME ENVIRONMENT SIMPLE AS THAT. TRACTION IS NOT $M_1V_1 = M_2V_2$ (ROCKET IMPULSE) BUT MORE LIKE YOUR FOOT PUSHING AGAINST THE GROUND COMPARISON. BECAUSE THE IONIC ATTRACTION IS LIKE A LOADED SPRING EXTERNAL GASES NEAR THE HULL SURFACE CANT GET OUT OF THE WAY FAST ENOUGH EVEN IF THE ATMOSPHERIC ENVIRONMENT WIND IS SUPERSONIC PAST THE HULL SURFACE, (SAUCER MOVING SUPERSONIC) - OH YES, DRAG REDUCES BECAUSE OF THIS PLASMATIC IONIC FLOW NEAR THE HULL. DONT LET MY WEAK PLASTIC SAUCER FOOL YOU, WITH GLASS THE VOLTAGE CAN BE RAISED ALOT HIGHER AND THERE IS A STATIC THRUST INCREASE WITH RISE IN VOLTAGE - QUITE HIGH FOR GLASS.

Document Scan - Faraday Law Saucer

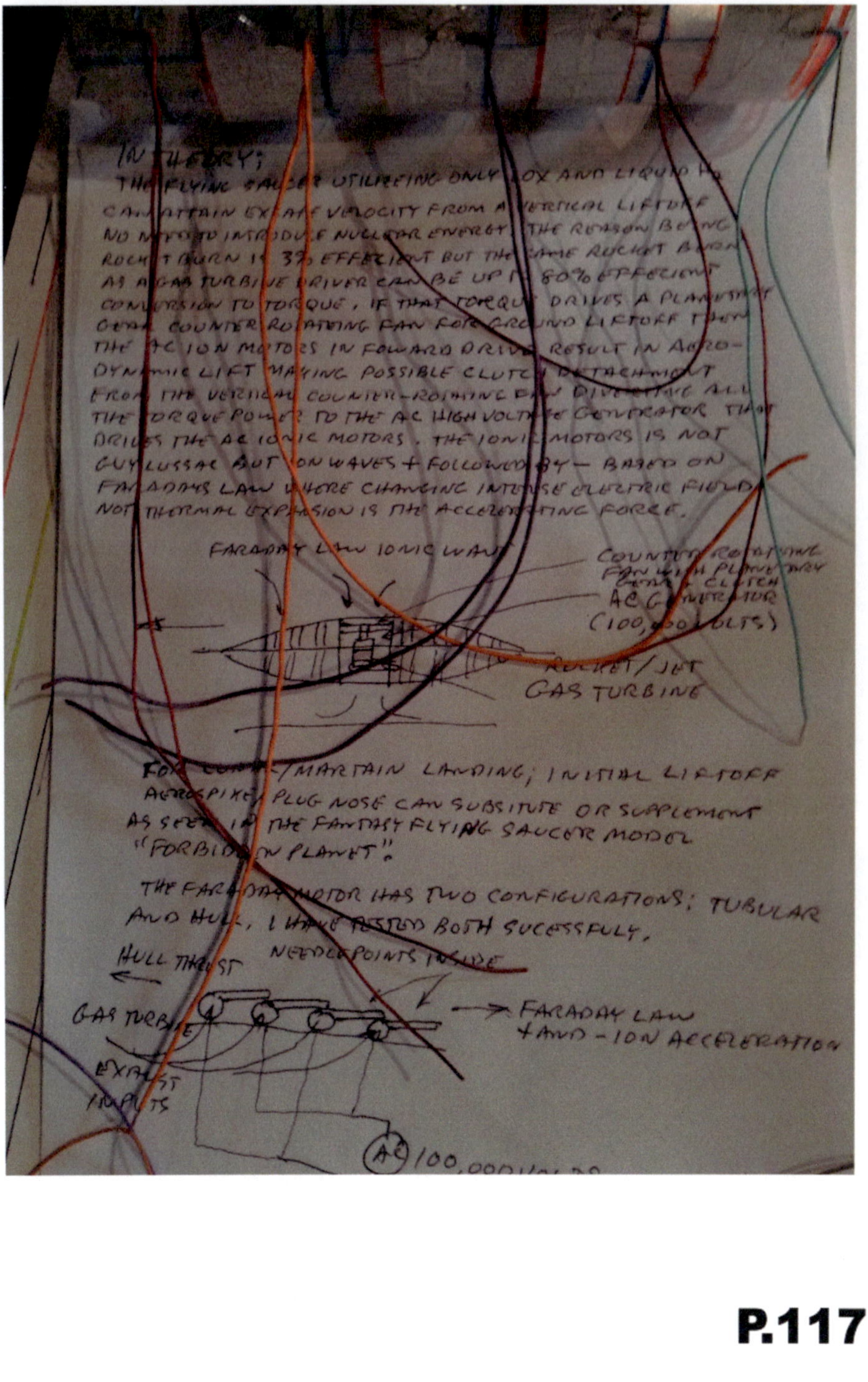

IN THEORY;
THE FLYING SAUCER UTILIZEING ONLY LOX AND LIQUID H2 CAN ATTAIN ESCAPE VELOCITY FROM A VERTICAL LIFTOFF NO NEED TO INTRODUCE NUCLEAR ENERGY, THE REASON BEING ROCKET BURN IS 3% EFFECIENT BUT THE SAME ROCKET BURN AS A GAS TURBINE DRIVER CAN BE UP TO 80% EFFECIENT CONVERSION TO TORQUE, IF THAT TORQUE DRIVES A PLANETARY GEAR COUNTER ROTATING FAN FOR GROUND LIFTOFF THEN THE AC ION MOTORS IN FOWARD DRIVE RESULT IN AERO-DYNAMIC LIFT MAKING POSSIBLE CLUTCH DETACHMENT FROM THE VERTICAL COUNTER-ROTATING FAN DIVERTING ALL THE TORQUE POWER TO THE AC HIGH VOLTAGE GENERATOR THAT DRIVES THE AC IONIC MOTORS. THE IONIC MOTORS IS NOT GUY LUSSAC BUT ION WAVES + FOLLOWED BY — BASED ON FARADAYS LAW WHERE CHANGING INTENSE ELECTRIC FIELD NOT THERMAL EXPANSION IS THE ACCELERATING FORCE.

FOR LUNAR/MARTAIN LANDING; INITIAL LIFTOFF AEROSPIKE/PLUG NOSE CAN SUBSITUTE OR SUPPLEMENT AS SEEN IN THE FANTASY FLYING SAUCER MODEL "FORBIDDEN PLANET".

THE FARADAY MOTOR HAS TWO CONFIGURATIONS; TUBULAR AND HULL, I HAVE TESTED BOTH SUCESSFULY.

Document Scan - American Flag Electronic Symbol

HI STEPHEN:

July 13, 2013

THIS I DISCOVERED BY ACCIDENT AFTER I DREAMED ALIENS IN ANOTHER GALAXY HAVE AN ELECTRONIC SYMBOL THAT LOOKS SIMILAR TO THE UNITED STATES FLAG (BY CO-INCIDENCE)

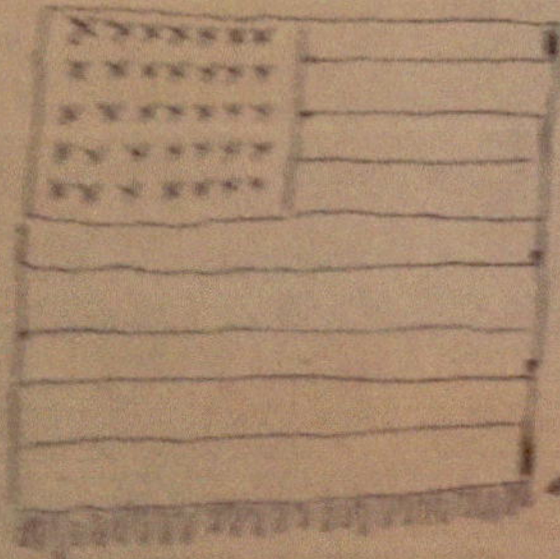

(STARS) STARS ARE MADE OF PLASMA. PLASMA RESPONDS TO ELECTRIC AND MAGNETIC FIELD. IN THE DARK MY NEEDLEPOINTS ARRANGED IN ARRAY CREATE TINY PLASMA POINTS IN THE PRESENCE OF AN INTENSE ELECTRIC FIELD.

BLUE: GAS AMBIENT

RED: AC INSULATED STRIPS

WHITE: SPACEING BETWEEN STRIPS

← THE MARINE FLAG HAS THIS TO PUT THE NEEDLE POINTS IN ARRAY. (BY-COINCIDENCE)

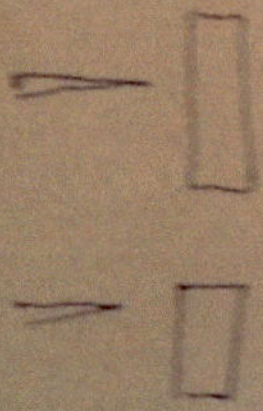

IF BOTH STRIP AND NEEDLEPOINT IS NUDE, THEN POSITIVE AND NEGATIVE IONS FLOW IN OPPOSITE DIRECTIONS RESULTING IN AN ARC - NO THRUST

IF THE NEEDLEPOINT IS NUDE AND THE METAL STRIP IS INSULATED - IONS CAN ONLY MOVE IN ONE DIRECTION. A THRUST IS GENERATED

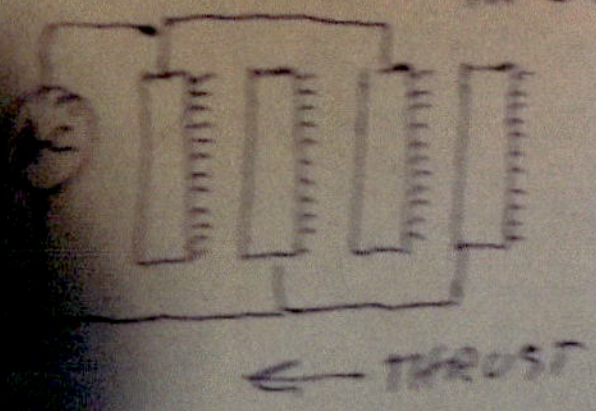

TO INCREASE THRUST: USE A HIGH PUNCTURE RESISTANT MATERIAL LIKE UNBREAKABLE GLASS - I DONT HAVE THAT SO I USE PLASTIC TO DEMONSTRATE THE MOTOR WORKS. WITH HIGH PUNCTURE STRENGTH GLASS THE VOLTAGE CAN BE INCREASED; INTENSIFYING THE FLOW

THE FREQUENCY IS INCREASED, THE NEEDLEPOINTS DONT NEED TO PROTRUDE RESULTING IN A SMOOTH SURFACE. IF THE NEEDLEPOINTS ARE IN TINY HOLES A GAS CAN BE PUMPED IN AND THE KIND OF GAS RESULTS IN A COLOR OR GLOW. THATS WHY UFOS CAN CHANGE COLOR. WHY AC? ITS EASIER TO BUILD LIGHTWEIGHT POWERFUL AC GENERATORS ELIMINATING ALOT OF HEADACHES ASSOCIATED WITH DC THRUST MOTORS.

Your Daddy
Gene Watson

Simple Index

1970 (year), 1, 6
1975 (year), 58
1970s (year), 58
1977 (year), 58, 72
1998 (year), 59

A

AC, 31, 32, 33, 35, 45, 48, 49, 56, 70, 76
aerospike, 6, 31
alpha particle, 30
Angels, 68
arc, 48
Arkansas, 1, 55
array, 7
Atlantis, 40
automatic valve, 53

B

B-field, 12, 15
ball, 71, 73, 74, 75
ball (ice), 23, 27, 28
band, 32
Bell Helicopter, 58
bevatron, 13, 16, 17, 14, 15, 21, 22, 27
65
blueshift, 38
Boltzmann's Constant, 23
boron glass, 9, 10, 12, 14, 21, 24
Dr. Bronco, 1, 6, 55
brushless (DC) motor, 53, 56
buoyant, 36
burn (cascade), 8
burn (chemical), 60
burn (hydrogen rich), 6, 72
burn (staged), 6

C

capacitors, 43
carnot cycle, 75
centrifugal force, 2, 14, 23
Chinese fan, 41, 42
coil (magnetic), 22
cold fusion, 71, 73, 74
compressor (staged axil flow), 5
counter rotating fan, 2
cyclotron, 11, 47

D

DC, 32, 43, 53, 56, 69, 72
Dees, 21, 27, 28, 45
delta wing, 44
deuterium, 23, 29, 63, 74
diodes, 32
directional transmission, 65
Drake equation, 65
dragonfly wings, 41
dream, 27, 69, 71

E

Einstein, 37, 39
electric field, 33
electric motor, 42
electron, 12, 15, 38, 65
electron gun, 14, 24
electron microscope, 74

F

FAA, 1
fail safe design, 11
fans, 41
Fan-Jet, 18
Fan-Jet-Rocket, 18
Faraday motor, 44, 45
Faraday's law, 33
filaments, 50
fission, 63

Simple Index

flapping, 41
Flash Gordon, 65
flywheel, 10, 14
foam glass, 31
"Forbidden Planet", 33, 76
frequency counter, 50

G

gas, 10, 16
gear reduction, 1, 2, 69
generalized sketch, 59
generator (AC), 31, 70
geometric configuration, 4
gravity, 64
gyroscopic, 2, 57

H

hadronic, 38
helicopter, 46
helium, 36
Higgs field, 39
high voltage, 72, 76
hydrogen rich burn, 6, 72
hypersonic, 7
hull, 45, 76

I

IC chips, 50
ice, 23, 27, 28, 63
ion, 19, 20, 29, 30, 32, 33, 34, 45, 43, 46, 48, 76
insulated, 30, 34
interstellar spaceship, 23
ITER, 62, 66

J

jet, 31, 57
Jet Rocket, 20

L

laser, 25, 27, 28
laser bank, 23
leptonic, 38
lithium batteries, 41, 69
lithium fusion, 25
Lorentz (radical), 39
low gravity, 27, 28

M

magnetic (field), 13, 21, 22, 28, 38, 55, 65, 74
Marconi radio, 66
mercury, 75
microcircuit, 50
microwave, 51
mirrored, 25
moon, 60

N

Native American, 67
needle array, 53
needle points, 30, 31, 35
negative ion, 43
nuclear battery, 30
nuclear fusion, 25, 62, 73

O

Ohm's Law, 55
oscilloscope, 49

P

Simple Index

particle accelerator, 9, 74, 75
pinch effect (magnetic), 55
planetary gear, 33, 57
plasma, 13, 19, 20, 22, 28, 29, 37, 75
Plasma Ionic Motor, 19
plastic, 36
positron, 25, 38
power, 49
propellant mass, 64
proton, 10, 12, 13, 14, 15, 22, 38, 64
proton beam, 24
pyramid, 40

Q

quarks, 75

R

radar, 51
reactor, 29, 73
red shift, 37
rest mass, 39
rocket, 5, 7, 9, 20, 31, 60
Rocket-Plasma-Ionic, 19
Rubbia, Carlo, 58
laser bank, 23

S

semiconductor, 30
SETI, 65
Skylon, 57
spaceplane, 4
spins, 46
spring down, 75
stack (magnet), 13, 47
static thrust, 16
Steady State, 37
string, 38
strips, 45
symmetrical, 13

T

Texas
- Grand Prairie, 58
- Hurst, 58
- Mountain Creek Lake, 58
- Sachse, 57
- Wylie, 2, 43, 32, 56, 70

toroid (donut), 14, 16, 17, 21, 22 , 25, 27, 28, 29
torque, 33
traction, 31
transformer (step-up), 43
tritium, 63
tube (double), 75
tubular, 44, 45
tungsten alloy, 24

U

US Space Command, 68

V

vacuum, 31
velocity (higher), 8
Vought Aerospace, 58

W

Wang Computer, 1
Watt, James, 38, 53
wind bike, 54
wings, 42

X

X-rays, 12, 52, 55, 60

References

Russellville Arkansas Polytechnic College: The Agricola 1970 378 AR48. Page 34 Charles Bronco, PhD Associate Professor Physics

Arkansas Polytechnic College 1971 Volum 55 Page 24, Charles Bronco, Assoc. Professor Physics

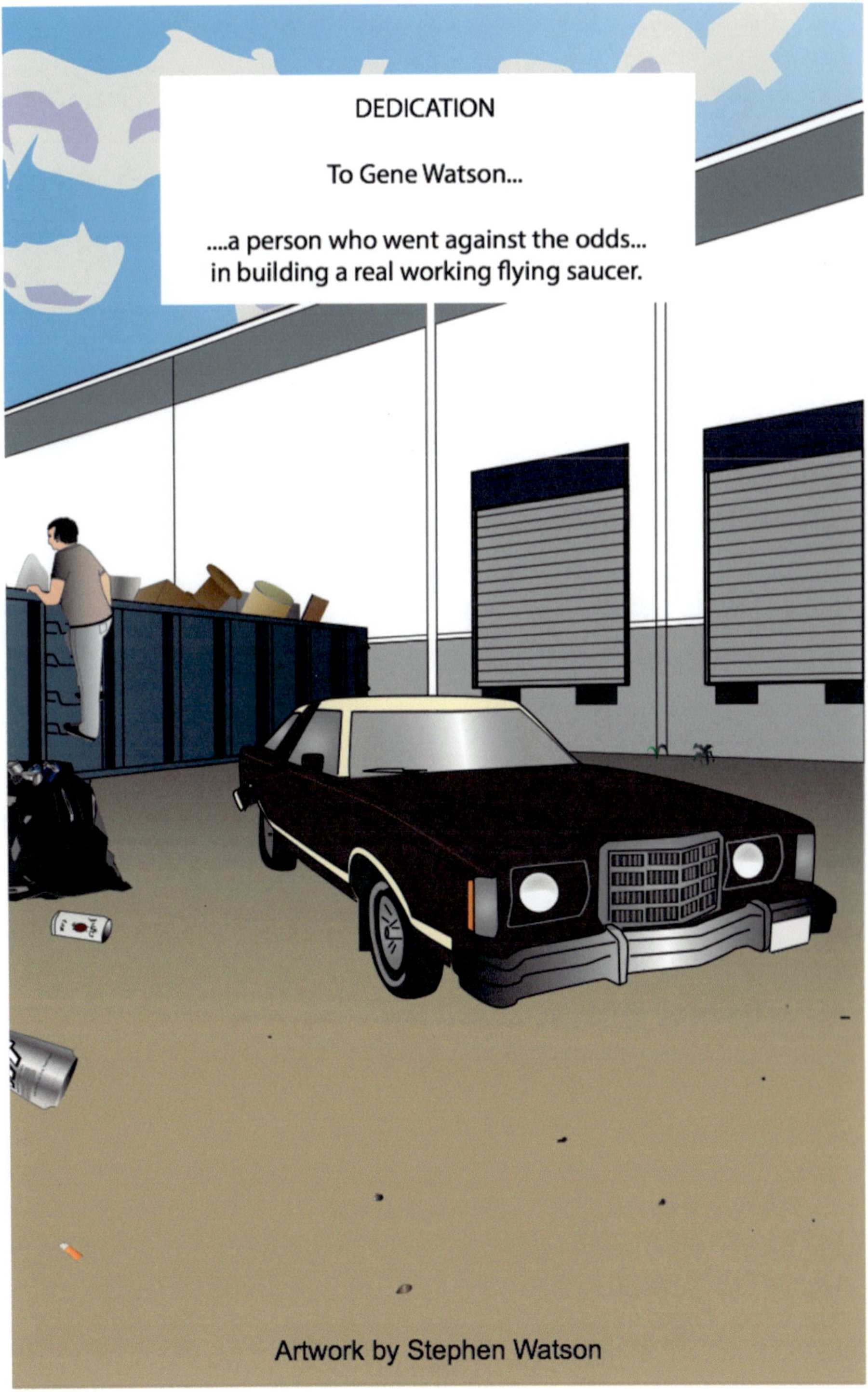
DEDICATION
To Gene Watson...
....a person who went against the odds...
in building a real working flying saucer.
Artwork by Stephen Watson

Made in the USA
Las Vegas, NV
07 November 2026

57673302R00081